T5-AFS-384

Biotelemetry II

Proceedings of the Second International Symposium on Biotelemetry
Davos, May 20–24, 1974

Biotelemetry II

Editor: *P.A. Neukomm,* Zürich

157 figures and 17 tables, 1974

S. Karger · Basel · München · Paris · London · New York · Sydney

QH 324
I 683 B
1974

LIBRARY
CORNELL UNIVERSITY
MEDICAL COLLEGE
NEW YORK CITY

NOV 6 1975

S. Karger · Basel · München · Paris · London · New York · Sydney
Arnold-Böcklin-Strasse 25, CH–4011 Basel (Switzerland)

All rights, including that of translation into other languages, reserved.
Photomechanic reproduction (photocopy, microcopy) of this book or part of it without special permission of the publishers is prohibited.

© Copyright 1974 by S. Karger AG, Basel
Printed in Switzerland by F. Reinhardt AG, Basel
ISBN 3–8055–2103–0

Contents

Session III. Storage Telemetry
Chairmen: *A. Shah and F. Pellandini*

Session IV. Preprocessing and Reduction of Telemetric Data
Chairmen: *G. Dumermuth and A.A. Borbély*

Session V. Telemetry of Biomechanical Parameters
Chairmen: *B.M. Nigg and M. Hebbelinck*

Session VI. Telemetry of Respiratory and Cardiovascular Parameters

Chairmen: *H. Hutten, Th.B. Fryer, H. Howald and H.P. Kimmich*

Session VII. Telemetry of Neurobiological Parameters

Chairmen: *A.A. Borbély, G. Dumermuth and J.A.J. Klijn*

Session VIII. Patient Monitoring – Clinical Telemetry

Chairmen: *H.P. Kimmich, T. Furukawa and J.A. Vos*

Congress Data

Organization

International Society on Biotelemetry (ISOB)

Swiss Federal Institutes of Technology, Lausanne (EPFL) and Zürich (ETHZ)
Institut für Technische Physik, Zürich (Head: Prof. Dr. E. Baumann)
Institut de Microtechnique, Lausanne (Head: Prof. Dr. C.W. Burckhardt)
Kurse für Turnen und Sport, Zürich (Head: Prof. Dr. J. Wartenweiler)

Organizing Committee

Chairmen
P.A. Neukomm, dipl. El. Ing. ETH, Laboratory of Biomechanics, ETH, Zürich
H.P. Kimmich, Dr. Ir., Department of Physiology, University of Nijmegen, Nijmegen

Secretaries
Mrs. M. Lebert and Mrs. V. Neukomm, Department of Physical Education and Sport, ETH, Zürich

Members
A.A. Borbély, Dr. med., Department of Pharmacology, University of Zürich, Zürich
C.W. Burckhardt, Prof. Dr., Department of Micro-Engineering, EPFL, St-Sulpice
H.U. Debrunner, Dr. med., Department of Experimental Orthopedics, University of Bern, Bern
G. Dumermuth, Dr. med., Department of EEG, Children's Hospital, University of Zürich, Zürich
Th.B. Fryer, Assistant Chief, Electronic Research, NASA, Moffett Field, Calif.
T. Furukawa, Prof. Dr., Research Institute of Applied Electricity, Hokkaido University, Sapporo
H. Howald, Dr. med., Research Institute, Federal School of Physical Education, Magglingen
H. Hutten, Prof. Dr., Department of Physiology, University of Mainz, Mainz
R. Jenzer, Jr. Ing., Exhibition Manager, Geroldswil
P.L. Kaefer, Dr., Press and Information, ETH, Zürich
H.P. Kaegi, Dr. med., Ärztefunkverein, Zürich
J.A.J. Klijn, Dr., Department of Neurophysiology, University of Nijmegen, Nijmegen
H. Kunz, dipl. El. Ing., Department of Applied Physics, ETH, Zürich
G. Matsumoto, Prof. Dr., Research Institute of Applied Electricity, Hokkaido University, Sapporo
J.D. Meindl. Dr., Prof. of Electrical Engineering, Stanford University, Stanford, Calif.
B.M. Nigg, dipl. Phys. ETH, Laboratory of Biomechanics, ETH, Zürich
F. Pellandini, Prof. Dr., Department of Mirco Technology, University of Neuchâtel, Neuchâtel
S. Salmons, Dr., Department of Anatomy, Medical School, University of Birmingham, Birmingham
M. Schoenenberger, Treasurer, Revisionsdienst ETH, Zürich
A. Shah, Dr., Department of Applied Physics, ETH, Zürich
J.A. Vos, Dr., Department of Physiology, University of Nijmegen, Nijmegen

Sponsoring and Cooperating Societies

Association Suisse de Microtechnique (ASMT)
Associazione Italiana di Ingegneria Medica e Biologica (AIIMB)
Biomechanics Laboratory, The Pennsylvania State University (Organizer of the IVth International Seminar on Biomechanics)
Deutsche Gesellschaft für Biomedizinische Technik e.V.
European Sleep Research Society
Fédération Internationale d'Education Physique (FIEP)
Fédération Internationale de Médecine Sportive (FIMS)
IEEE Engineering in Medicine and Biology Group
IEEE Switzerland Section
International Committee on Physical Fitness Research (ICPFR)
International Council of Sport and Physical Education (ICSPE-CIEPS)
The International Society on Biomechanics
The Research Committee of ICSPE-UNESCO
International Council on Health, Physical Education and Recreation (ICHPER)
Schweizerische Arbeitsgemeinschaft für Biomedizinische Technik (SABT)
Schweizerische Vereinigung für Elektroenzephalographie und Klinische Neurophysiologie
Stanford Electronics Laboratories, Stanford University, Calif.
Université de Neuchâtel, Département de Microtechnique

Sponsoring Companies

CIBA-GEIGY AG, Basel, Switzerland
W.W. Fischer, Morges, Switzerland
F. Hellige & Co. GmbH, Freiburg/Breisgau, FRG
Kistler Instrumente AG, Winterthur, Switzerland
Sandoz AG, Basel, Switzerland
SFENA S.A., Velizy-Villacoublay, France
Stiftung Hasler-Werke, Bern, Switzerland

Contributors

Cynar SA, Zürich
Kambly AG, Biscuitsfabrik, Trübschachen
Gebr. Kümin, Weine und Spirituosen, Freienbach
Lateltin AG, Spirituosen, Zürich
Lindt & Sprüngli AG, Chocoladenfabriken, Kilchberg
Martini & Rossi SA, Zürich
H. Nobs & Cie AG, Nährmittelfabrik, Münchenbuchsee
Produktion AG Meilen, Lebensmittelfabrik, Meilen
Roland Murten AG, Zwieback-, Sticks- und Knäckebrotwerke, Murten
Zweifel Pomy-Chips AG, Spreitenbach

Introduction

Biotelemetry is the realistic measurement of physical and chemical parameters from a living mobile subject and its surroundings. The technique employed can vary from storage telemetry to radio frequency transmission links. Biotelemetry has established its usefulness in both research and clinical practice wherever signals must be measured accurately with maximum freedom of movement.

These proceedings of the 2nd International Symposium on Biotelemetry held in Davos, May 20–24, 1974, represent technical developments and further applications which have evolved since the first symposium (Nijmegen, 1971). The contributions come from a variety of disciplines including biomedical engineering, applied physics, physiology, orthopedics, neurobiology, biomechanics, behavior research, sport, physical education, patient monitoring and many others.

In order to publish the proceedings as soon as possible, all manuscripts are printed as written by the authors, with only minor changes. I extend my thanks to *F. Pellandini* and *R. V. Mazza* for their efficient reviewing of the manuscripts during the conference.

At Davos there was appreciable discussion among the symposium participants coming from 24 countries, and this interaction was valuable to all those present. These proceedings by no means represent the complete work of the authors, but rather are intended to give the reader an overall view of the current state of biotelemetry. The reader is encouraged to personally contact the authors for additional details.

I wish to thank the committee members for their support, the authors for their contributions and all those who participated in the 2nd International Symposium on Biotelemetry.

Zürich, August 1974 *Peter A. Neukomm*

Session I

Telemetric Equipment and Transducers

Chairmen: *C.W. Burckhardt and P.A. Neukomm*

Biotelemetry II. 2nd Int. Symp., Davos 1974, pp. 2–4 (Karger, Basel 1974)

Personal PDM/PCM Biotelemetry System

H.P. Kimmich and H.J.B. Ijsenbrandt
Department of Physiology, University of Nijmegen, Nijmegen

Introduction

With the development of an oxygen uptake ($\dot{V}O_2$) biotelemetry system for use under certain circumstances (Kimmich and Kreuzer, 1972) there has been an increasing interest in monitoring this parameter also under different and more general circumstances with better defined and possibly greater accuracy.

We are at present trying to improve the principle of dynamic $\dot{V}O_2$ measurement by several measures, namely design of a new flow transducer for use over the whole physiological range from rest to maximal exercise incorporating an ultrasonic principle, theoretical investigation of the accuracy of the dynamic $\dot{V}O_2$ measurement (preliminary communication, Kimmich, 1974), electronic improvement of the response time of the PO_2 transducer (Kimmich, Kreuzer and Hoofd, 1973), accurate fixation of the delay introduced by the PO_2 measuring system and reduction of the error introduced by the electronic delay network (realized but not published), and improvement of the transmission quality by use of pulse code modulation (PCM).

The application of C-Mos integrated circuits with their voltage flexibility and low power requirements make the design of complex circuitry, as is required for PCM telemetry, feasible. A major design requirement was easy conversion of a PDM system presented previously (IJsenbrandt and Kimmich, 1972) by simply interchanging the coding or decoding print respectively.

Materials and method

The description of the engineering aspects of the telemetry system is restricted to the coding (submodulation) and decoding system since the remaining parts have been described earlier. The 50 mW transmitter radiating in the 150 MHz band is constructed (similarly to the receiver) of a cabinet, housing the coder and RF transmitter, and up to 7 interchangeable preamplifiers. By connecting it to a battery and voltage stabilizing unit, the latter is activated supplying the full battery voltage of 7 to 10 volts to the transmitter and the stabilized voltage of + and -2.5 volt to the preamplifiers and transmitting unit. The transmitter (10cm x 7cm x 2 cm) including batteries and preamplifiers weighs 280 g. Such a system has the advantage of easy matching of size and weight of the supply unit with the requirements of duration of the experiment. In addition remote control of the power switch is possible, by simply adding a unit between transmitter and battery unit, with the aid of telecontrol.

A block diagram of the submodulator is shown in figure 1. The

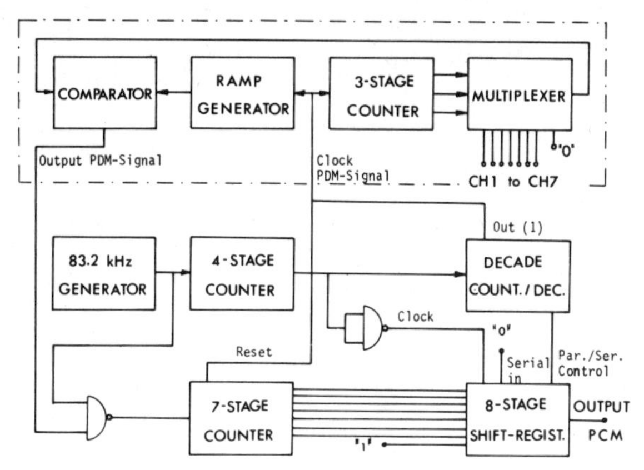

Figure 1

The submodulator of a PDM/PCM telemetry transmitter is composed of 7 C-Mos integrated circuits and 20 discrete elements. It is composed of two prints, one for the generation of the PDM signal, and the other for conversion to PCM. Current supply is 1.2 mA.

analog signals (one being zero for synchronization purposes) are multiplexed with the aid of an 8 channel multiplexer (CD 4051 AE). The binary control data are obtained from a three stage counter which in its turn is triggered by each 160th pulse of the 83.2 kHz clock generator of the PCM system. (For exclusive PDM operation this 520 Hz clock is generated directly). The pulse duration of the 520 Hz clock is then modulated with the multiplexed analog signal using the well-known method of a ramp generator and a comparator (modulation m= 0.05 to 0.95).

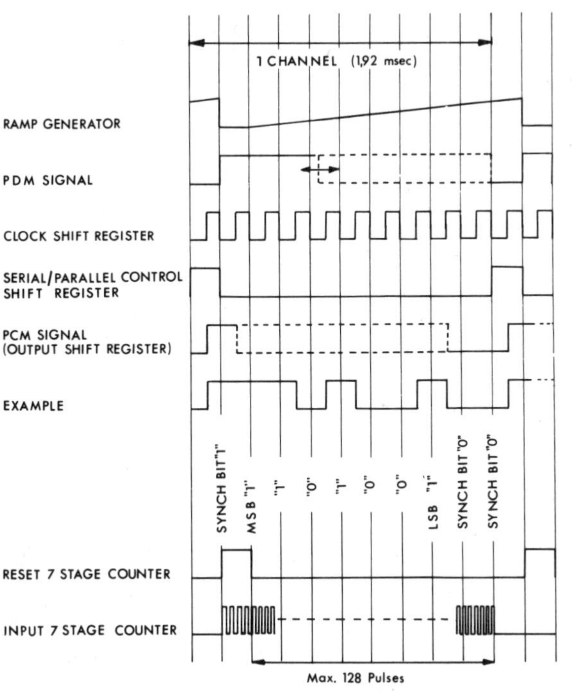

Transfer of the pulses of the 83.2 kHz clock to a 7 stage binary counter (corr. more for more than 7 bit operation) is controlled by the duration of the individual pulses of the PDM signal. The first 16 pulses of maximal 144 pulses (2^7+ 16) are not counted because they coincide with the reset of the counter (fig. 2). After completion of the count the information is parallel-shifted to the shift register (the last input being a logic one). The information is then serial-shifted by each 16th pulse of the 83.2 kHz

Figure 2

Wave forms of a PCM channel (see text)

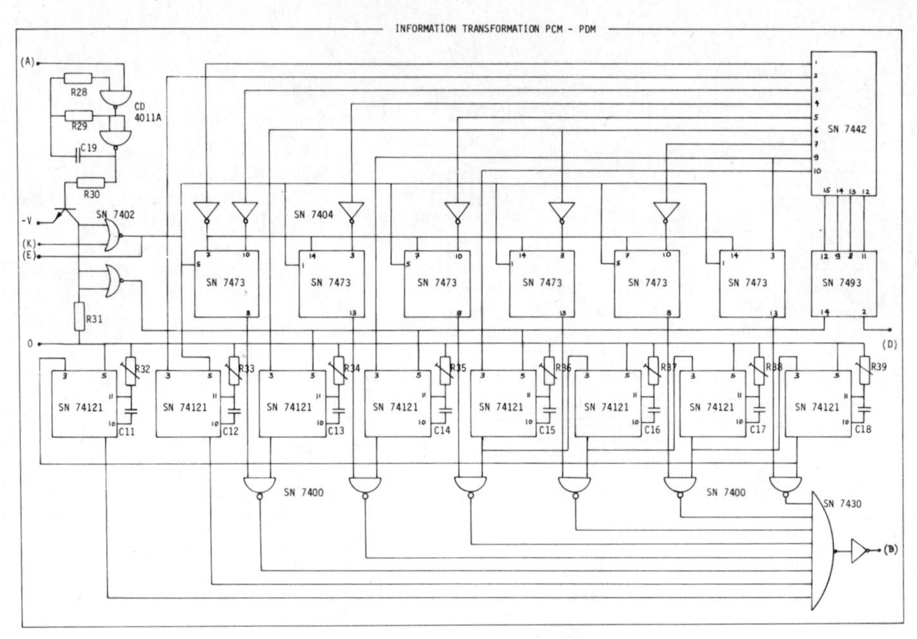

Figure 3: Information transformation PCM to PDM at the receiver.

clock (5.2 kHz). At the same time the counter counts the pulses of the following channel after being reset. The PCM signal obtained at the last output of the shift register is composed of a synchronization bit (logic 1), 7 data bits, and again two synchronization bits (logic 0).

At the receiving side the data is converted from PCM to PDM (fig. 3). This is slightly more complicated than simply incorporating a D/A converter. However, in such a way only a single print has to be interchanged, which practically is done by a switch (PCM-PDM control). Each data bit triggers a one-shot with well defined shot time according to its value (from MSB to LSB). One synchronization bit also triggers a one-shot in order to assure safe operation also with missing data on some channels and to have proper DC calibration between PCM and PDM.

Conclusions

With this method a simple conversion of PDM systems to PCM is possible giving all the well-known advantages of PCM, such as greater accuracy and direct compatibility to digital data processing at the receiving side.

References:

KIMMICH,H.P., (1974) CLINICAL TELEMETRY AND PATIENT MONITORING. This edition.

KIMMICH,H.P. and KREUZER,F.(1972) TELEMETRIC DETERMINATION OF OXYGEN UPTAKE DURING EXERCISE. Int. J. Biomed. Engng., 1/1 (27-36)

KIMMICH,H.P., KREUZER,F. and HOOFD,L.J.C.(1973), Digest of the 10th Int. Conf. Med. Biol. Engng. Dresden, 40/6 (146)

IJSENBRANDT,H.J.B. and KIMMICH,H.P. (1971) In: BIOTELEMETRY, Ed.: H. P. Kimmich and J. A. Vos, (57-64), Meander, NV, Leiden

Biotelemetry II. 2nd Int. Symp., Davos 1974, pp. 5–8 (Karger, Basel 1974)

Criteria and their Measurements in Multichannel Biotelemetry

Horst Kaltschmidt
MBB, München

Contrary to the single-channel telemetering equipment, the multi-channel equipment permits an investigation of biological systems by using the well proven tools of cybernetics, particularly information and control theory. The fundamental priniciples of cybernetic investigation methods consist of finding correlations of two or several system functions and their mathematical description. The simplest system (see Fig. 1) contains only one input and one output that are accessible by means of measuring technique and that are more or less causally connected. A well known practical example for the use of Multi-Channel Biotelemetry is the medical investigation of sporting men, where the values of ECG, breath, acceleration, temperature, etc. have to be transmitted. Let us also refer to the already used automatic ECG-evaluation methods which need at least three ECG Signals in synchronism with the heart beat. These few examples of applied cybernetic methods already show the scientific demand for multi-channel telemetering equipment. The definition as well as the measurability of the quality criteria for these multi-channel telemetering systems is of the same importance. Fig. 7 shows the basic concept of a multi-channel telemetering system.
The criteria referring to the mobile station (bord system) are the following: number of channels, bandwidth, voltage range, input impedance, maximum voltage out of signal bandwidth, stability of power supply for transducers with defection bridges, safety factor against multi-channel propagation, the distance between bord equipment and ground equipment, and finally shock and vibration, and temperature range as well as weight, volume, and operating time. The criteria for the receiver and hence for the whole system, too, are - measured at the output - the signal-to-noise ratio, the linearity, the crosstalk of the individual channels with each other, the transfer function, the various time delays.

As far as the individual concepts are not self-evident, such as weight and volume, their definitions shall be given by means of the test circuits which define them. The difficulty of quantifying or measuring the criteria consists in that a number of influence factors, which are not easily kept constant, act on the quantities to be determined. Above all,the board-ground distance as well as the environmental noise should be mentioned here. If the environmental noise is very considerable, the measurements can be made in a screened chamber (EMC-measurment chamber). The distance between transmitter and receiver are simulated by an inserted damping link. The influence of the distance between transmitter and receiver physically results in a reduction of the received power thus causing an increasing signal-to-noise ratio at the input of the receiver, and hence the signal-to-noise ratios of the individual test channels gradually deteriorate more and more. Fortunately, somewhat more sophisticated modulation methods, such as the FM and PCM, are used in most telemetering systems. They offer considerable noise suppression until a threshold. These methods are especially characterised in that they offer a good signal-to-noise ratio up to a certain threshold value (for a certain distance), but suddenly give a very bad signal-to-noise ratio beyond this threshold. A number of test circuits shall follow now in the same sequence as they are used in practice for determining the quality criteria (Fig. 9 and the following). As the following pictures together with the legends, contain enough information regarding the methods of measurment,no further text is given.

The question of which multi channel biotelemetry is the best one depends on many criteria with respect to the application. As these criteria can be measured,an objective comparsion between various telemetry systems can be made.

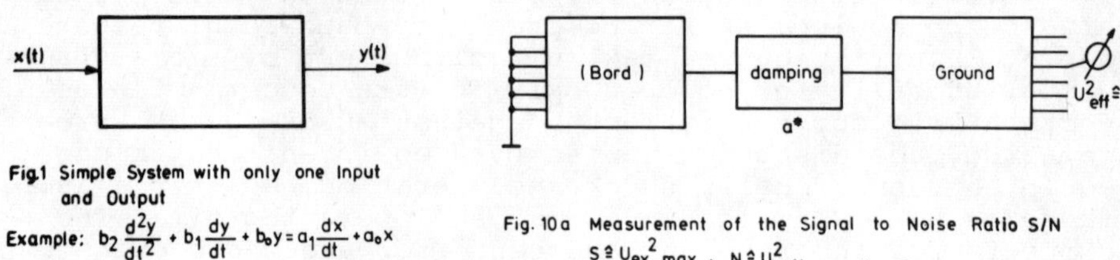

Fig.1 Simple System with only one Input and Output

Example: $b_2 \frac{d^2y}{dt^2} + b_1 \frac{dy}{dt} + b_0 y = a_1 \frac{dx}{dt} + a_0 x$ (linear)

$b_2 \frac{d^2y}{dt^2} + b_1 (\frac{dy}{dt})^2 + b_0 y^2 = a_1 \frac{dx}{dt} + a_1 x^{\frac{1}{2}}$ (non linear)

Fig. 10a Measurement of the Signal to Noise Ratio S/N
$S \hat{=} U_{ex}^2 \, max$, $N \hat{=} U^2_{eff} \, max, U_{in} = 0$

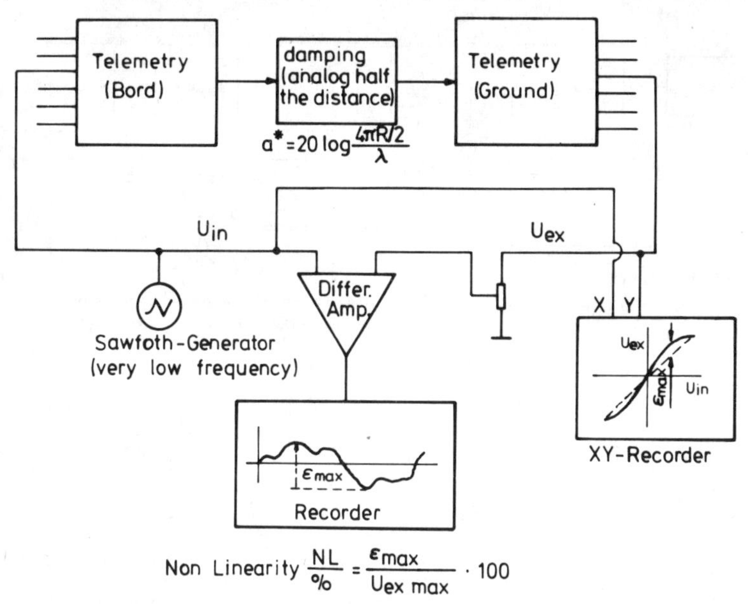

Fig. 9 Measurement of the Linearity by the differential Method

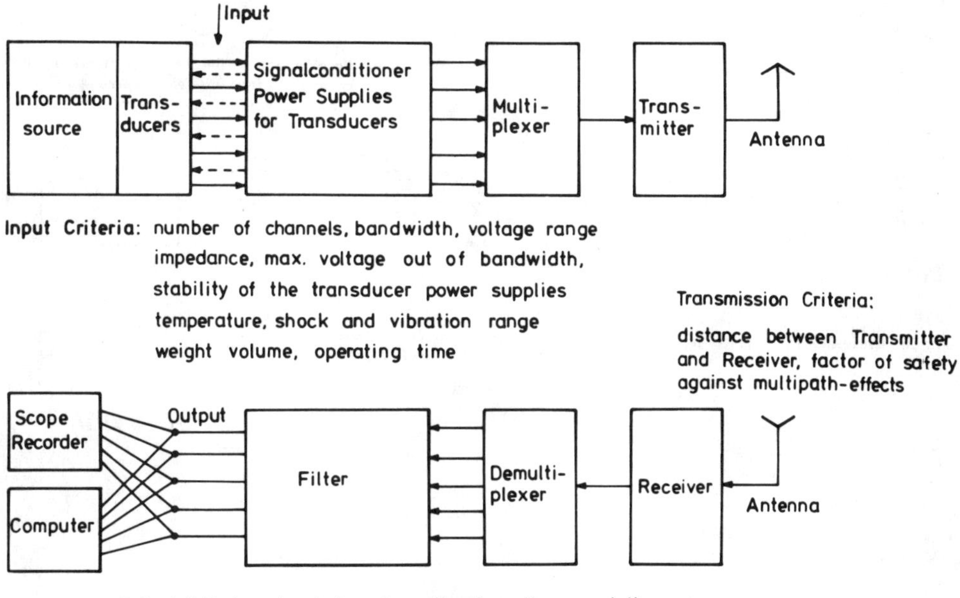

Output Criteria: signal to noise ratio, linearity, crosstalk transfer function, different time lags

Fig. 7 Basic Concept of a Multichannel Bio-Telemetry and its Criteria of Quality

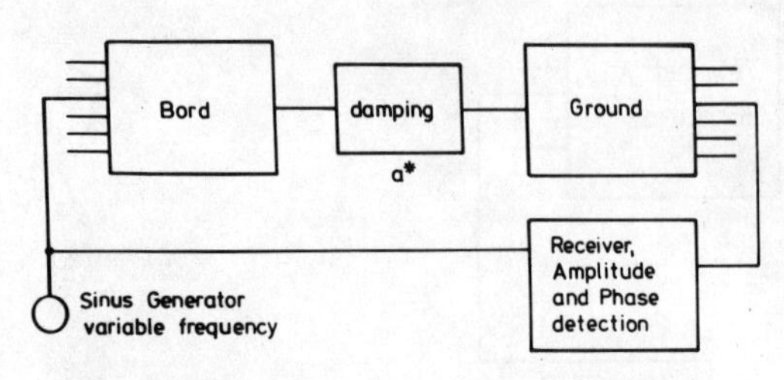

Fig. 12 Measurement of Amplitude and Phase Transfer Function

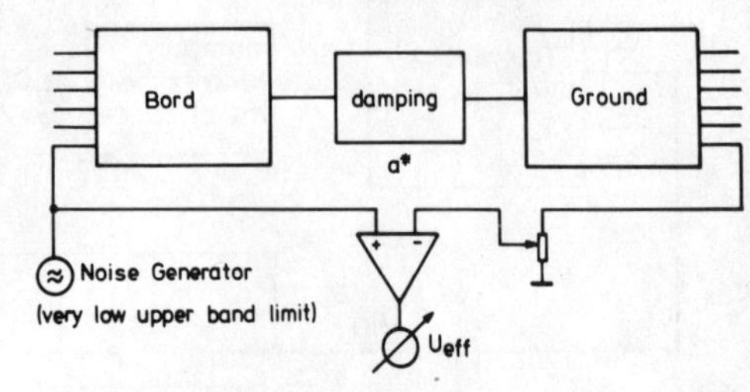

Fig. 11 Measurement of the Noise Component by the differential Method $N = U^2_{eff}$

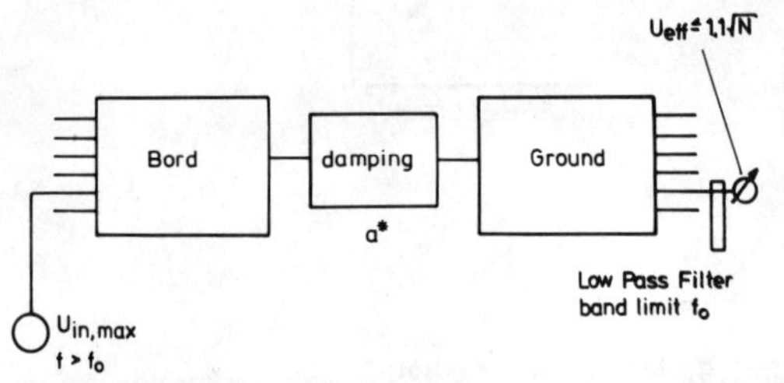

Fig. 14 Measurement of the maximum Input Voltage out of the bandwidth f_o

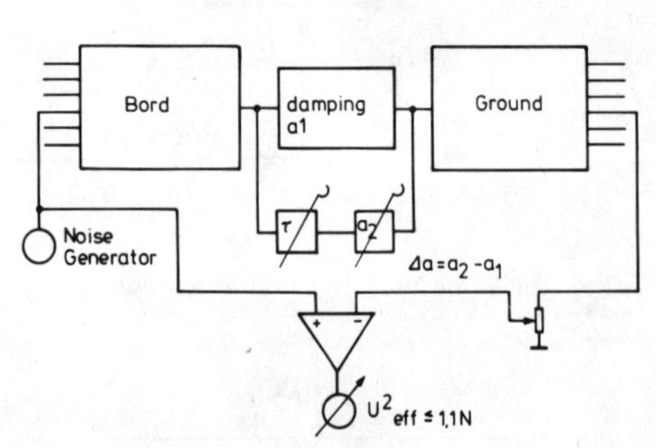

Fig. 16 Measurement of the factor of safety against multipath effects (FM-Telemetry $\tau = \frac{1}{4 f_{Hf}}$)

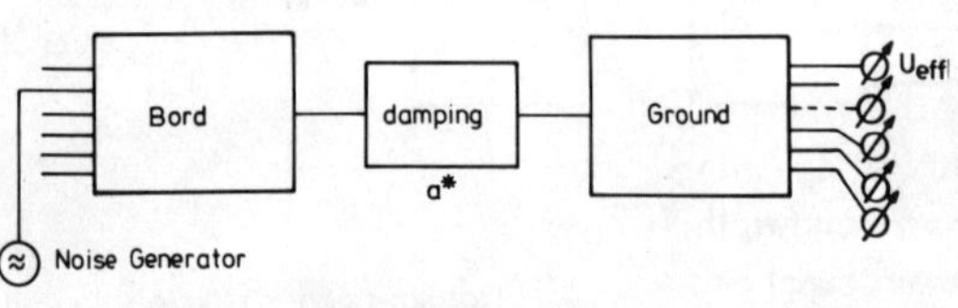

Fig 13. Measurement of Crosstalk

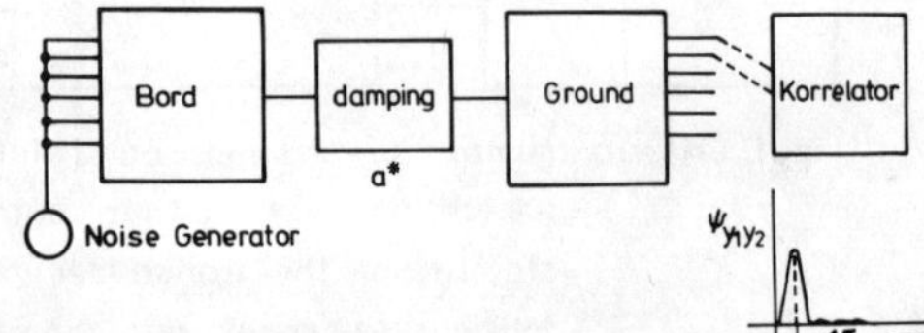

Fig. 15 Measurement of differential time lags by Correlator

Biotelemetry II. 2nd Int. Symp., Davos 1974, pp. 9–11 (Karger, Basel 1974)

ECG-Transmitter System for Good Battery Efficiency

Helmut Leist and Karl-Heinz Günther
Hellige GmbH, Freiburg/Breisgau

The design of a low range ECG-telemetry transmitter led us to the problem of selecting the optimum supply and battery voltages. High quality of signal transformation was to maintain, and sufficient output power was to generate, with good efficiency.

1. Battery voltage selection for RF-circuits.

1.1. Influence to RF-power: Choosing the battery voltage for a miniature transmitter defines the maximum RF-power that is to be generated with a certain simple circuitry. For bipolar transistor circuits it was shown in a lot of applications by other authors, for instance DEBOO et al. (1964), that the RF-power generated with the voltage of one mercury cell is sufficient for low range telemetry. We measured 2mW maximum RF-power at 36MHz into 50Ω with a 1,3V DC-supply. Increasing the battery voltage in the same circuit resulted in a linear increase in current and a quadratic increase in RF-power. Efficiency was nearly constant with a slight reduction below 2,5V due to saturation voltage.

1.2. Influence on efficiency from current variation: On the other hand a reduction in the output-stage current below 3mA at 1V supply resulted in a rapid decrease of efficiency. With the oscillator included we got 25% efficiency at 4mA supply current and 15% at 2mA for BSX20. This is due to a decrease of transit frequency at low current values. Because you can't reduce the current proportionally, you get, for a preselected low RF-power, a reduced efficiency at higher battery voltages. Conditions are here different from those of ROCHELLE (1974), who looks for the maximum power output, at low frequencies.

1.3. Number of cells: Thus for a given volume or weight and short range it is advantageous to choose one big cell for the battery instead of two or more smaller ones, regarding in addition: the current rating, the capacity or hours of operation, the ease of replacement, storage and costs.

2. Supply for LF-circuit: LF-circuits with bipolar silicon epitaxial planar transistors have U_{BE} of 0,5V at 10μA and U_{CEsat} of 0,1V, thus may also work down to a 1V supply regarding easy control and voltage swing. But only low common mode rejection up to 30 and a subcarrier oscillator linearity of ±2% for ±30% frequency deviation is possible. However, a CMR of 60dB or more is necessary, if the transmitter may be removed from the skin. Discrete differential amplifiers with current sources or integrated operational amplifiers require at least a 2V total supply for good CMR and output voltage swing. CMR was >80dB for a 20...200μA, 2V total supply to ICL8021C.

2.1. A regulated DC-DC-converter: With the small current consumption of LF-circuits it is possible to generate a second voltage with a DC-DC-converter. A blocking oscillator is useful, but it must be stabilized for varying battery voltages, as shown in Fig. 1:

Transistor T_3 works as a blocking oscillator with the applied transformer and C'. The base is fed by R' and T_2. Neglecting T_1 and R_3, the voltage divider R_1, R_2 senses the rectified and filtered negative output voltage and reduces the base current of T_3, if U_2 increases, thus producing a U_1/U_2 proportional to the resistor ratio. Adding T_1 and R_3, the temperature coefficient may be chosen following the formulae in Fig. 1 for the resulting influence of U_{BE}. The efficiency is 30% with a transformer of 7mm diameter pot core at 100KHz, - 1,5V output for 1V input and a load of 20kΩ, 45% for a 2kΩ load.

<u>2.2. A voltage regulator at 1V level:</u> If you also want to use rechargeable batteries and everywhere available zink-carbon types, it is necessary to stabilize battery voltage for constancy of the subcarrier oscillator and for calibration purposes. A simple method like that of KRAFT (1972) would not be sufficient here.

Fig.2 shows a circuit that works down to a 1,1V battery voltage. Neglecting T_1 and R_1, T_2 is the voltage sensor at the voltage divider R_3/R_4 and its base-emitter voltage gives the reference. T_3 is merely an inverter for driving the series regulator element T_4.

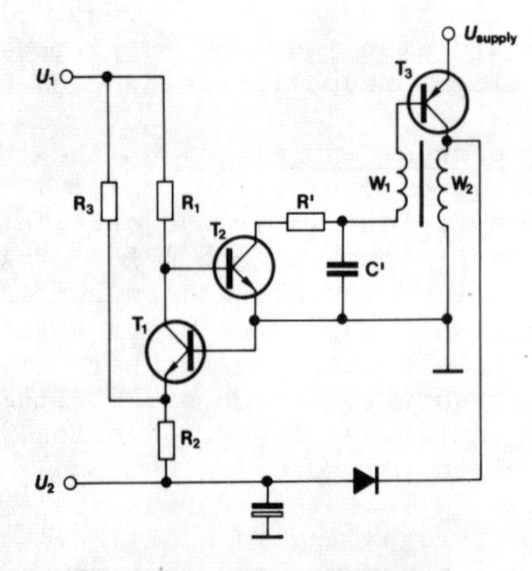

For $U_{BE2} = U_{BE1} = U_B$ and $I_{B1} = I_{B2} = 0$

$$\frac{U_2 - U_B}{R_2} = \frac{U_1 - U_B}{R_1} + \frac{U_1 + U_B}{R_3}$$

$T_c = -2{,}2\,\frac{mV}{K}$ for $R_1 = R_3$; $I_{C1} = I_{C2}$

$T_C = 0$ for $\frac{R_2}{R_3} = \frac{R_2}{R_1} - 1$; $I_{C1} = I_{C2}$

Fig.1: Regulated blocking oscillator for DC-DC conversion

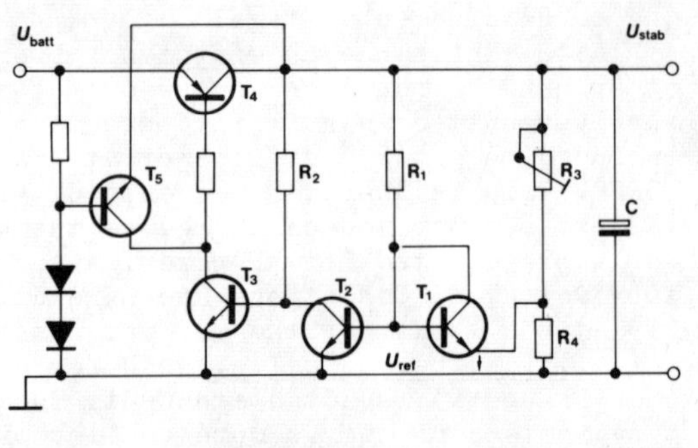

$U_{ref} = a\,U_T$; $a = \ln \frac{I_{C2}}{I_{C1}}$ $\frac{R_1}{R_2} \cdot \frac{I_{C2}}{I_{C1}}$

$T_C = 0$ for a $\Delta U_T = \frac{R_4}{R_1}\,\Delta U_{B2}$; $\frac{R_1}{R_4} = \frac{26}{a}$

$U_T = 26mV$ at 300K; thermal voltage

$dU_T/dT = 86\mu V/K$; $dU_{BE}/dT = -2{,}2mV/K$

Fig.2: Voltage regulator for 1V output and low input voltage

If R_2 is connected to the stabilized voltage instead of the unstabilized one, as might be possible, you increase stabilization factor from about 10 to more than 1000. Now a start-up transistor T_5 with a simple reference voltage is necessary. The output resistance is about 100Ω for a 2:1 load change and I_2=10μA. The circuit is stable for an output capacity C of more than 30nF.

To get a temperature coefficient other than -4,4mV/K you must add transistor T_1 and Resistor R_1. Now a base voltage differential of T_1 and T_2 is generated by different currents through the transistors choosing R_1 and R_2 appropriately. This differential voltage serves as a reference with a known positive temperature coefficient. This is compensated by a current variation through R_4 via R_1, T_1. That is produced in R_1 by the base-emitter voltage drift of T_2 for constant U_{stab}.

2.3. Subcarrier oscillator supply: The conventional oscillator is a symmetric astable multivibrator with current sources instead of the base resistors. Frequency modulation is achieved for equal voltages applied to the bases of both current source transistors. To reduce the loading of the negative supply the basic multivibrator is connected only to the stabilized positive voltage. The current sources are drift compensated by an equal temperature coefficient of the negative supply voltage for frequency stability.

3. Transmitter performance: The transmitter RF-stages of the unit are driven directly from the battery, while the LF-circuit is driven from the stabilized positive voltage and the negative stabilized voltage derived from it. The current consumption of the whole unit with crystal oscillator and separate RF-output stage is 6,7mA at 1,3V. 10% of this goes to the miniature DC-DC converter and to the remaining LF-circuitry. The battery voltage may vary within ±15%, function is to +200%. Range in the plane field with whip antenna of 50cm at the transmitter and matched $\lambda/4$ receiver antenna is 40m without dropouts for 1,2V battery voltage, the transmitter adapted to a standing person. The transmitter works 100 hours with a RM 401 battery, that is about Lady-size (IEC R1). The transmitter shows good performance with the indicated circuits, there is no disturbance from the blocking oscillator. Details about the system in production are given by DITTMAR (editorial 1974).

Acknowledgement: The authors wish to thank Mr.C.Behr and Mr.H. Lindblad, SRA, for the final realization of the telemetry system, also Miss Schill and Mrs.Blankenhorn for careful measurements, as well as Mr.G.Ullrich and Mr.P.Schlüssel, HELLIGE, for the motivation and helpful discussions.

References:

DEBOO,G.J; FRYER,T.B.: A miniature biopotential telemetry system NASA tech.brief 64-10171 (1964).
DITTMAR: (ed.) Biotelemetriesystem Hellige Meditel 37, Arbeitsmed. Sozialmed. Präventivmed. 9: 61-63 (1974).
KRAFT,W.: Leistungsarme Komplementärschaltungen für kleine Speisespannungen. Elektronik 21: 389-392 and 429-431 (1972).
ROCHELLE,J.M.: Design of gateable transmitter for acoustic telemetering tags. IEEE Trans. on Biomed. Eng. 21: 63-66 (1974)

Biotelemetry II. 2nd Int. Symp., Davos 1974, pp. 12–15 (Karger, Basel 1974)

Telemetric Measurement of Intracranial Pressure with the Help of an Electromagnetic Detector

G. Foroglou, R. Favre, R. Besse and E. Zander

Service de Neurochirurgie et Division Autonome d'Electronique Médicale, Hôpital Cantonal Universitaire, Lausanne

The telemetric intracranial pressure recorder developed in our hospital consists basically of an implanted detector which is additionally influenced by the alternating magnetic field of an externally placed inductor coil.

The detector (Fig. 1) is made up of a mobile tubular magnet (3) placed along a guide ruby crystal (2). The latter is fixed to the base (5) of the cap "shaped" capsule (1). The open aspect of the capsule is occluded by an air tight film of mylar a few µm thick. One pole of the magnet rests on the bottom of the capsule while the other lightly opposes the mylar diaphragm.

When correctly positioned in its operational mode (Fig. 2), the detector (1) is implanted within the cranial cavity (7). A secure fit is established by a form of bayonet fixation. Three ear-like projections (4) are incorporated into the wing of the "cap" and three corresponding gutters are cut into the margins of the implanting drill hole. Having introduced the unit, a slight rotation will bring the ear projections to bear firmly between the dura mater (8) and the cranial vault.

The tubular magnet is partly subjected to an axial hydrostatic force transmitted across the dura mater, which is proportional to the intracranial pressure, and partly to alternating fields of magnetic force (of 50 Hertz) generated by the externally positioned inductor unit (9) which also works axially. When the repulsive forces of the magnetic field exceed the hydrostatic force, the magnet develops a steady axial vibration, quite audible externally and very clearly through a stethoscope.

The exceptional qualities of $SmCo^5$ magnets guarantee a linear association between the intracranial pressure and the induced current, registered within audible range. Rapid fluctuations of pressure up to 5 Hertz of weak strength may be recorded by amplitude modulation of the trace they produce when the detector is slightly overcharged (Fig.3).

The lower portions (11) of the trace (10) detected by a special microphone, represents the variations of intracranial pressure. The inductors position is of prime importance in obtaining the best results. One should pay particular attention to the geometrical relationship of two variables : the relative alignment of inductor and detector axes, and the distance separating these units.

When first seeking a maximal signal from the detector, the inductor coil is held vertical to the skull over the implantation site. The orientation may be facilitated by the use of either a compass or a needle having a thread which will be attracted by the magnetic field of the detector.

The separation of inductor from detector is set according to geometrical considerations. A large inductor unit will tend to reduce any errors stemming from a malalignment.

Refined instruments embody a large inductor held by an articulated support. The incorporation of a magnetic field sensor in the center of the inductor head will greatly simplify its optimal operational placement in relation to the detector.

The detectors heads have an intrinsic margin of error of 1%. The collective influence of secondary factors may increase the error under operating conditions to the order of 10 to 20%, whereas with superior equipment in skilled hands, this becomes 2-3%.

An assessment of transdural pressure is considered practical by several workers provided that the dura mater has not been deformed.

We studied the role of the dura mater by simulating different aspects in the laboratory. Depending on its physical and geometrical properties, this intermediary membrane was seen to alter the active surface where the intracranial pressure, acting as a hydrostatic force, bears on the magnet. The changes are reflected in a simple rotation of the response graphs (Fig. 4) about a fixed point of intersection (14).

Knowing the calibration point (12) established by the manufacturers, the instruments response in clinical use is given by a single recording taken at the time of implantation (13) and made in reference to conventionally employed standards. This calibration point, along with the point of intersection (14), will provide the appropriate response graph (15).

This technique is subject to a relative variable in the nature of changes in atmospheric pressure which are said to be transmitted through to the interior of the detector capsule. An allowance was made for the effect by providing a number of perforations in the capsule head.

Investigations are underway to determine the latency of equilibration following alterations in atmopheric pressure, be it an ambient change or a rapid change in altitude in ambulant patients.

In principle, the implanted device has an unlimited working life, being unaffected by ageing or changes in temperature. This is qualified, of course, by certain factors of material quality of everything that concerns the diaphragm and its fixation.

The stainless steel 18/8 originally used for the detector capsules was bitterly criticised for its residual magnetism, very much a function of the casting which distorted their smooth function. We have replaced it by an alliaz of bronze and glucynium (Glucydur) less subject to oxydation and having the additional strength sought for the bayonnet projection.

Despite the apparent simplicity of the detector units, the very fine precision demanded for the guide filament which is necessary to prevent a secondary lateral vibration (a play of 2 µm) in addition to production problems posed by $SmCo^5$ magnets, brings their cost price to Fr. 100-200 depending on the execution.

Trials on physiological tolerance are currently being run in rabbits, before advancing on to operational application to humans.

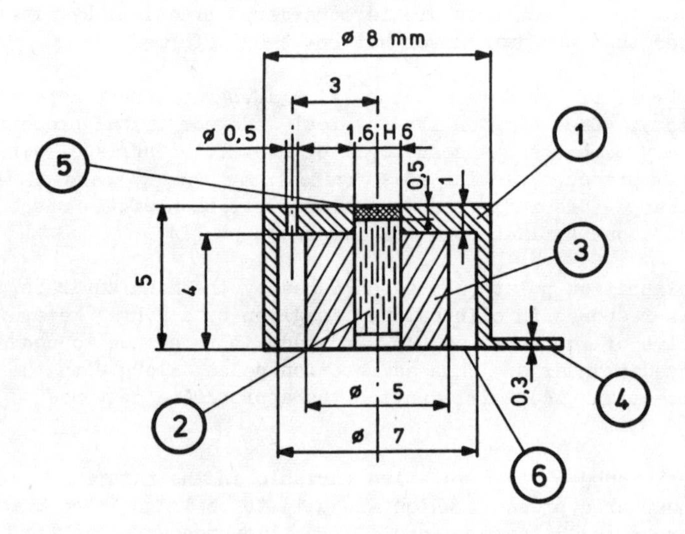

Fig. 1 : Axial section of the detector.

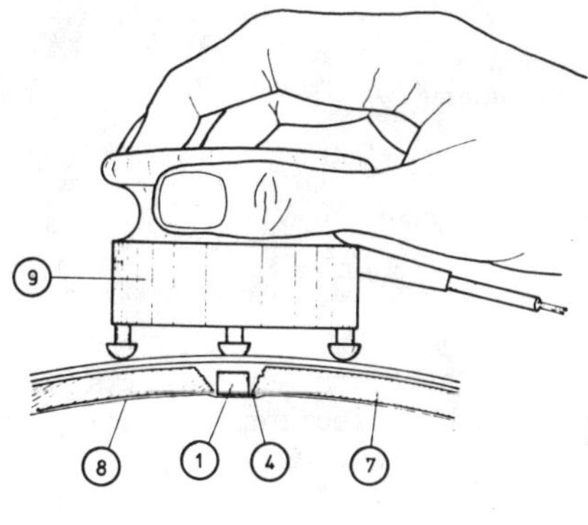

Fig. 2 : Detector in its operational position

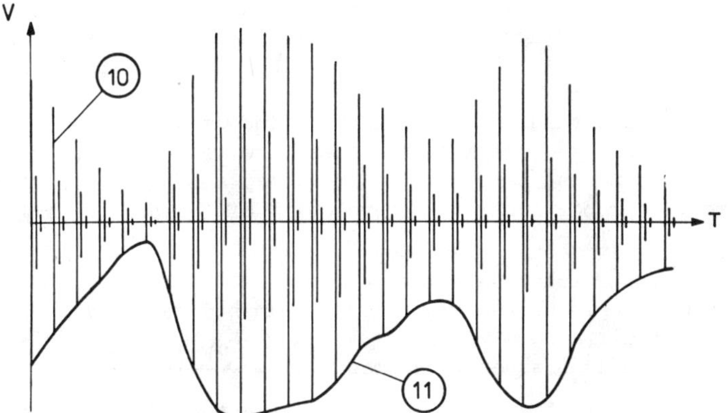

Fig. 3: Measurement of rapid fluctuations and at low levels of the intracranial pressure.

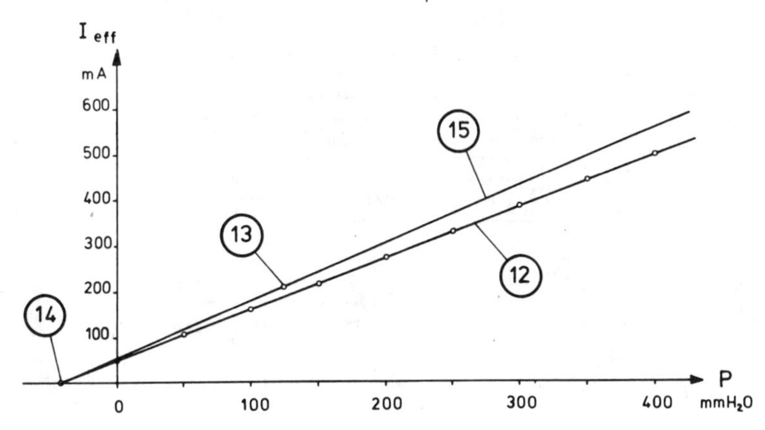

Fig. 4 : Influence of the dura mater on the detector's response.

Biotelemetry II. 2nd Int. Symp., Davos 1974, pp. 16–18 (Karger, Basel 1974)

An Implantable Directional Doppler Flowmeter

D. Cathignol, C. Fourcade and J. Descotes
INSERM – Chirurgie vasculaire et Transplantation d'Organes, Bron

Long term monitoring of blood flow after the experimental transplantation of an organ is impaired by infections and tolerance problems arising from the connection heads which are used to join the transducer and the measurement unit. In order to overcome these problems, we have developed an implantable flowmeter.
To help understanding of the modifications which are involved, we recall in fig. 1 the principle of the standard directional Doppler flowmeter.
A piezoelectric transducer emits a continuous wave ultrasonic beam where direction is angular toward the vessel axis. Ultrasound reflected by moving particles is frequency shifted: this variation is directly proportional to the velocity of the particles when:

F : ultrasonic frequency
V : particles velocity
α : angle between ultrasound beam and velocity
C : ultrasound velocity

The variation in frequency becomes : $DF = \dfrac{2F\ V\ \cos\alpha}{C}$

DF is positive if the particles are moving toward the transducer, and negative if they move backward.

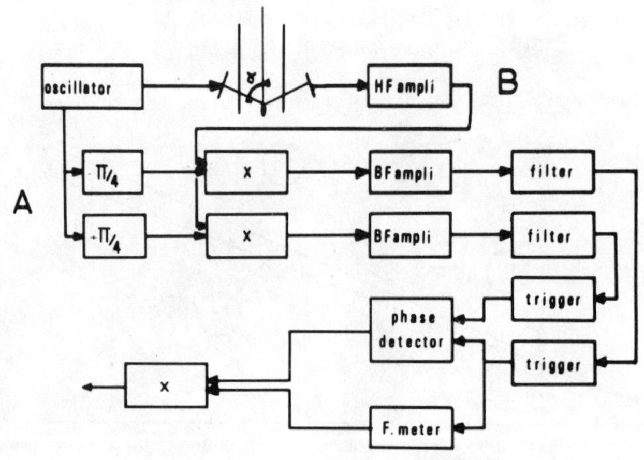

Figure 1. LOGIC INTERCONNECTION of DIRECTIONAL DOPPLER FLOWMETER

Mc LEOD was the first to provide evidence that DF magnitude and sign detection is possible.
The schematic diagram on fig. 1 is inspired by his work.
The signal obtained from the receiving transducer is amplified, then sent to two mixers together with the reference tension coming from the oscillator: this tension is phase shifted of $+\pi/4$ on one channel and $-\pi/4$ on the other.
At the output of the mixers, we obtain two low frequency signals phase shifted of + or $-\pi/2$ according to DF being positive or negative.
A phase detector and a frequency meter display the magnitude and sense of the frequency variation.
In order to transmit the algebric value of F one would have to implant the HG oscillator, the HG amplifier, the low frequency signal analysis and the 100 MHz emitter. This can't be considered as major problems of size and energy consumption would appear.
Thus we have considerably reduced the implantable part of the device: it only includes the oscillator which provides excitation for the transducer and the HG amplifier which is also used for transmission of the signal.
Oscillator: we saw that for the further treatment of the signal a reference is needed. As it does not exist any more in our system, we shall have to reconstitute it. Any variation in the instantaneous phase between the new reference and the transmitted HF signal will result in noise; both oscillators must thus have the greatest stability possible which compels us to use quartz oscillators.
HF amplifier: the signal obtained from the second transducer is amplified then transmitted to an imperfect antenna made from a steel plate.
Reception: we must reconstitute the reference; this is obtained by a phase lock with a quartz variable oscillator for the above mentioned reasons.
An automatic gain control regulates the amplitude of the received signal and makes the system distance-free till about 50 cm.

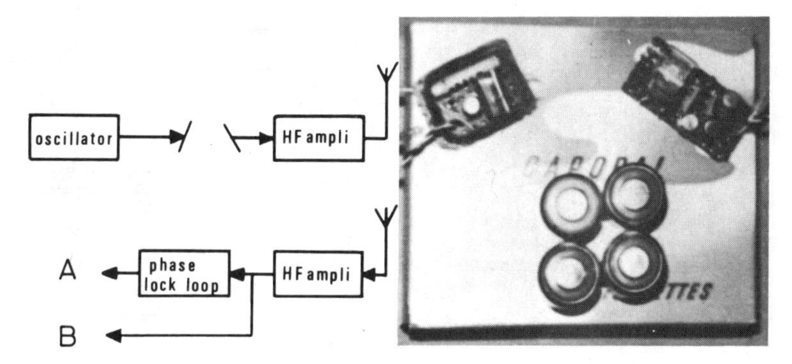

Figure 2. Logic interconnection of implantable flowmeter and different parts of this flowmeter.

Transducers: Perivascular transducers are usually made either from acrylic resins or from polyurethan. We found several disadvantages with them for long term implantation:

-their rigidity leads to arterial wall necrosis through continuous friction,

-the teflon or PVC coaxial wires do not allow perfect long term tightness between the transducer and the implantable flowmeter. We have been realizing our transducer and connection leads from medical grade silastic; the flowmeter is itself encorpulated in this plastic and thus the whole system is water-proof and perfectly well tolerated.

Prototype and results: the prototype is a rectangular bloc whose dimensions are 5x3x2cm, in order to avoid consumption, the flowmeter is activated only during the measurements. Autonomy allows 1 000 measurements of 1 minute.

We show on figure 3 recordings obtained from the descending aorta, the transducer being placed in one sense and then the other, with a short time of clamping of the alerty downward in both cases.

Conclusion: In opposite to the flow transmitter sensitive only to the moduler of the velocity (FRANKLIN, POURCELOT) and the phase shift technique (Rade), the system described here offers the following advantage:

-stable zero, no control after implanting the device, no HF.

We have been using the same system to transmit the intravascular bubbles found in the ultrasonore detector described elsewhere.

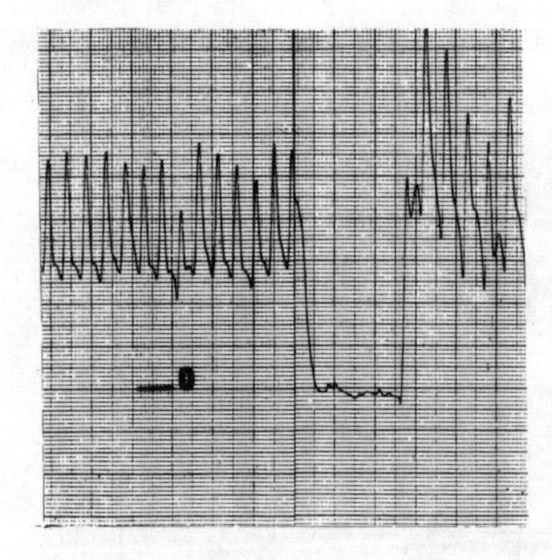

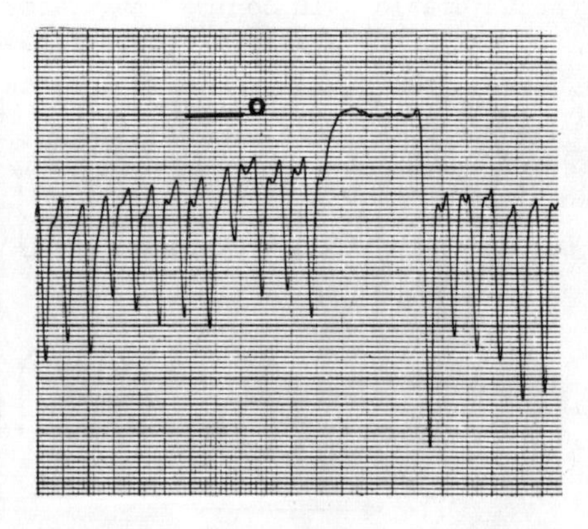

Figure 3. Aortic flow: the probe is faced in one sense and in the other sense.

Biotelemetry II. 2nd Int. Symp., Davos 1974, pp. 19–21 (Karger, Basel 1974)

Jaw Position Telemetry for the Control of Mobility Aids

Joachim A. Maass and Vera S. Maass

US Army Materiel Command, Frankfurt/Main, and University of Missouri, Kansas City, Mo.

A mobility aid should increase the number of options for independent activities available to a handicapped user. A wheel chair which should qualify as a mobility aid for quadriplegics must be motor driven and fully controllable by its user.

Mobility aids must meet a number of requirements shall the damages not offset the gains. These requirements can be summarized as follows:

* The device should contribute to the user's independence without adding to his burden.
* It must be safe, reliable, simple to use and as versatile as possible.
* Most important: It must be acceptable to the user.

Existing designs often violate one or more of these requirements. Myoelectric potentials as control inputs require the skillful application of galvanic electrodes with all their disadvantages. If the potentials are derived from facial muscles, electrodes and wiring will disfigure the user. Direct contacts between user and electronics imply reduced safety. Requirements on processing electronics and user training would be formidable. Hence, such a device would violate nearly all of the above requirements. Eye-movement detectors are even less favorable. Apart from the fact that they also disfigure the user, they interfere with precisely those sensory functions which are needed to navigate the wheel chair. They also fatigue delicate muscles which are actuators for small saccadic and pursuit movements but are far too weak for the bang-bang ocular rotation required for a reasonably reliable operation of an eye-movement-controlled wheel chair. In general one can say that a biological information input should not be used as an output for control purposes.

The device which is introduced in this paper is based on a preliminary study of user needs and a proposal for satisfying the stated requirements (1).

The proposed system

* does not require special training for its use or application,
* conserves the user's energy,
* allows for full visual control of its function,
* is practically invisible,
* does not require electrical connections between user and equipment.

The system is based on the telemetry of voluntary maxillary movements. The relative positions of the jaws are monitored by means of passive

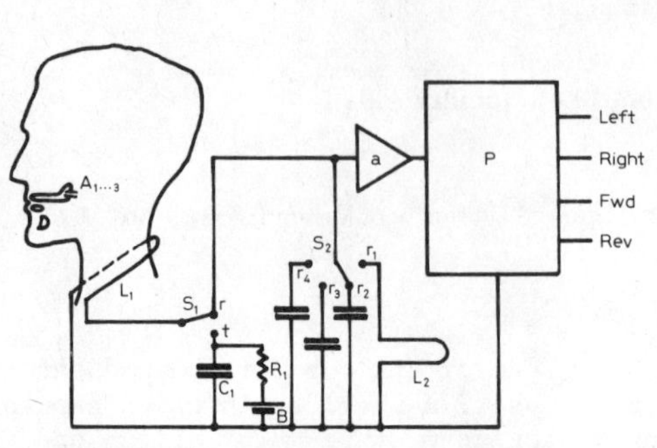

Figure 1, Schematic Diagram

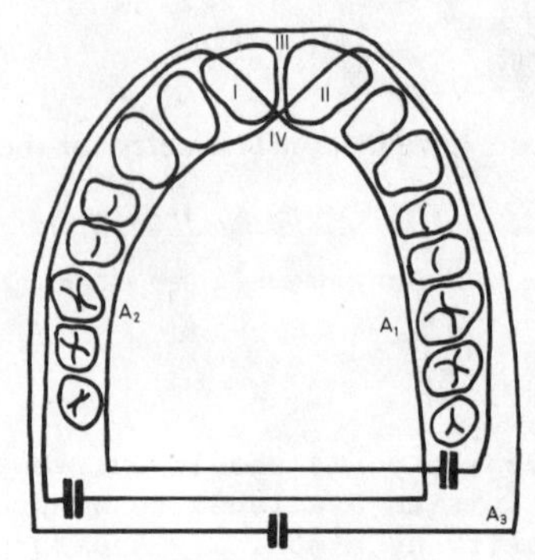

Figure 2, Resonant Circuits

FM-transponder circuits which are worn in the mouth and whose resonance frequencies are interrogated via a loop antenna which is worn around the neck and easily concealed by the user's garments.

In the schematic diagram (Fig. 1) A_1, A_2 and A_3 are resonant circuits of which one is indicated. They are built into a cosmetic splint which fits over the natural teeth of the upper jaw or they are contained in a dental prosthesis. The resonant frequencies of the three circuits can be influenced by a guitable inductively coupled detuning device D,e.g., a closed-loop conductor, which is concealed in a second splint, attached to the lower jaw.

Switch S_1 alternates between positions t and r respectively. In position r the battery B charges capacitor C_1 through resistor R_1. In position t the capacitor is discharged through the antenna loop L_1. The inductively coupled resonance circuits $A_{1...3}$ are triggered into damped oscillations whose frequencies are received when S_1 is at r. Frequency selector switch S_2 tunes L_1 to one of the A-frequencies in a repeating sequence from r_1 to r_4 which is synchronized with S_1 by a clock in the processor P. The processor which follows the amplifier, a, demodulates the FM and derives control signals for the wheel-chair motions left, right, forward and reverse.

Position r_1 of S_2 is used to monitor environmental noise. Excessive interference would turn off the equipment and/or trigger an alarm. Of the various possible configurations for the A-circuits, one is shown in Figure 2. The detuning device can be situated opposite the extremum positions I through IV or it can occupy any intermediate position with "no movement" corresponding to a center position. For detection of relative jaw positions in the horizontal plane alone, circuits A_1 and A_2 would suffice. However, circuit A_3 can be used to detect opening of the mouth, e.g. in yawning or talking. The processing electronics could be made to incorporate memory which would retain the last command until proper operation is restored. Certain patterns of relative vertical movements of the jaws could also be used for additional control commands such as turning on and off the equipment.

In an actual circuit it probably would be advisable to accomplish the tuning by the lock-in of a phase-locked-loop (PLL) circuit whose oscillator frequency could be FM-demodulated by a pulse-counting detector.

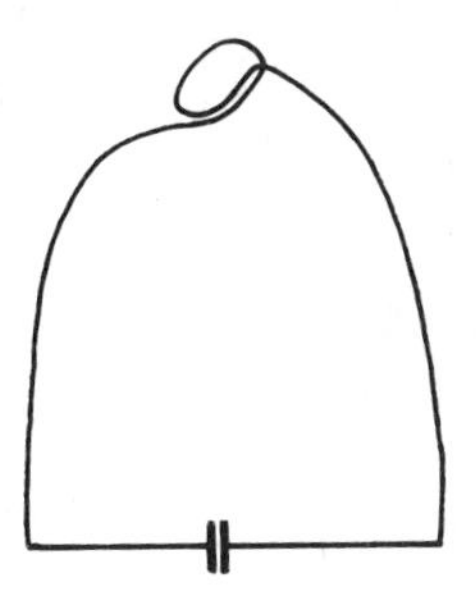

Figure 3,
A-circuits with concentrated inductances

Proportional control signals could then be easily derived from pulse counts during a clock-gated interval or by comparison with a standard frequency. This type of digital processing would also allow for a rather simple implementation of a dead zone around the "no-move" position.

If the resonant circuits are operated in the 13.56 MHz-band, proposed for medical applications (2), the inductances could have the form of single-wire loops around the teeth, tuned by capacitors of approximately 4 nF. A detuning loop of approximately 10 mm diameter would produce a frequency shift in the order of maximally one kilohertz. This can be increased by increasing the loop diameter or by introducing concentrated inductances in the A-circuits (Fig. 3). The narrow-band operation is favorable for a high signal-with-noise/noise ratio.

The frequency stability of the resonant circuits is high due to the thermostatic action of the oral cavity. A high mechanical stability must be assured since changes in the geometry of the circuits would change their resonant frequencies.

User's preference as to which jaw is to indicate the direction and speed of motion can be accommodated simply by a polarity reversing switch. All of the transmitter, receiver and processing electronics are contained in a box which is worn in a pocket or built into the wheel chair. They can be readily implemented from available integrated circuits. Production automation should pose no serious problems. The only exception being the customer-fitted splints.

Additional applications for the device can be found in all areas where it is desirable to free the hands of a machine operator for more important tasks while certain two-dimensional motions could be servo controlled by jaw position telemetry.

1) MAASS J.A. and MAASS V.S.: Mobility aid for quadriplegics, Jackson and DeVore (Eds.) Proc. 1973 Carnahan Conf.Electrnc. Prosthetics, Nov. 1973, 123 - 125 (Univ. of Kentucky).

2) KIMMICH H.P. and VOS J.A.: Biotelemetry, Post office regulations of some European countries concerning the use of radio waves for application to (medical) telemetry, 418 (Maender N.V., Leiden, 1972).

Biotelemetry II. 2nd Int. Symp., Davos 1974, pp. 22–24 (Karger, Basel 1974)

Studies to Compensate Temperature Effects in Measurements of Respiratory Oxygen

Georg Küchler, Wolfgang Wagner, Rudolf Schneiderreit and Ilse Wolburg
Zentralinstitut für Arbeitsmedizin der DDR, FB Arbeitsphysiologie, Berlin

Continuous measurements of oxygen consumption by means of radiotelemetry require compensation of temperature dependence of the polarographic P_{O_2} sensitive receptor. In previous investigations by KIMMICH[1] and in our own laboratory [3] this was done by measuring the oxygen partial pressure at a place where the temperature of the respiratory gas was constant, using a heating block. In general this method is a good one but there are some disadvantages:

- the delay increases between changes in the oxygen concentration of the respiratory gas within the mask and the response of the P_{O_2} receptor,
- mixing of different oxygen concentrations cannot be completely excluded on the way from the mask to the heating block,
- the amount of power supply needed for the heating block and for a pump is relatively high (about 2 watt) and for this reason the time available for telemetering measurement is limited.

The effect of changes in temperature on the output of the P_{O_2} sensitive receptor used was therefore investigated and compensation of the temperature effect by using thermistors was tried. First a thermistor was placed within the body of the receptor, forming part of a lattice network in the output circuit. Static characteristics of this arrangement are shown in fig. 1. The uncompensated output of the P_{O_2} receptor has a temperature dependence of about 4.0 Torr O_2 per ^{o}C [cp. 2]. The output controlled by thermistor varies in relation to the adjusted working resistance and shows undercompensation, compensation, and overcompensation respectively. It should be noted that the temperature characteristic of the thermistor used and that of the P_{O_2} sensitive receptor do not agree exactly with each other. The wider the range of changes in temperature the stronger is the divergence from the ideal compensation.

Besides this static behaviour the P_{O_2} receptor shows typical dynamic reactions to sudden jumps in temperature: a new output level is reached after several minutes, but at the very beginning, during the first second, the slope of the trace is steeper than afterwards. If a thermistor controlled P_{O_2} receptor is used, in the trace of the output this may be recognized as a small peak, of which the temperature dependence is about 0.5 Torr O_2 per ^{o}C. The appearance of this peak means that there is some delay between the effect of a change in temperature on the P_{O_2} receptor and on its

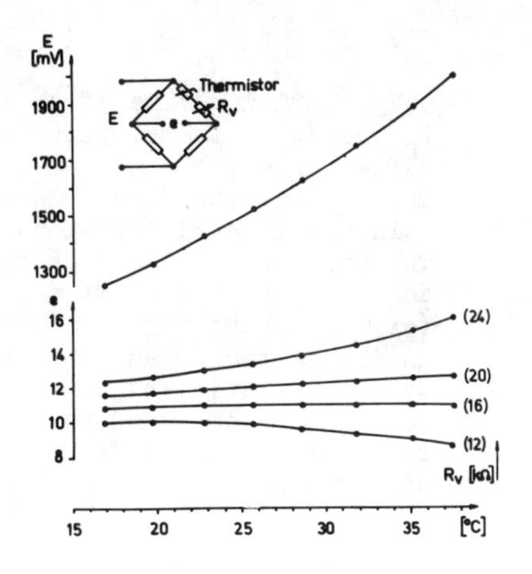

Fig. 1

Static temperature dependence of the polarographic P_{O2} sensitive receptor. upper curve: uncompensated P_{O2} receptor curves below: controlled by thermistor situated within the body of the P_{O2} receptor. Different working resistances (Rv) are adjusted.

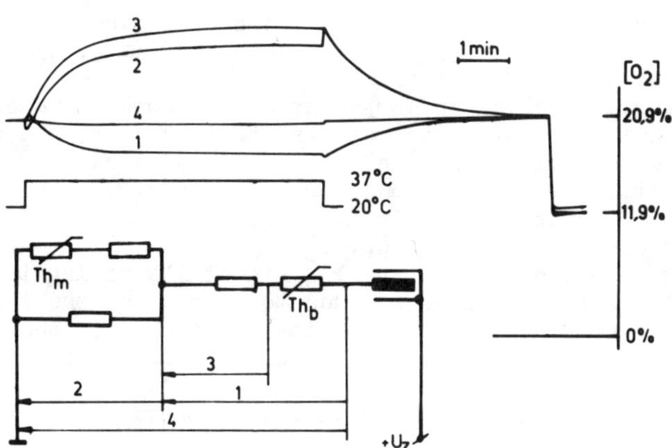

Fig. 2 Dynamic temperature dependence of the P_{O2} receptor controlled by 2 thermistors situated in the body of the receptor (Th_b) and close in front of the teflon membrane (Th_m). Simultaneous registration from 4 points of the compensation circuit during a jump of temperature from 20 °C to 37 °C. Amplifiers are so adjusted, that a jump in oxygen concentration effects in all 4 points a voltage deflection of the same amplitude; 1: curve of the slow thermistor; 2: curve of the fast thermistor; 3: corresponding to uncompensated behaviour; 4: compensated

controlling thermistor. To reduce this temperature-dependent peak a second thermistor was arranged close in front of the teflon membrane of the P_{O2} receptor. Registrations with this new arrangement were made by leading off several points of the compensation circuit during jumps of temperature (fig. 2). The temperature compensated curve shows a reduced peak, the temperature dependence is 0.15

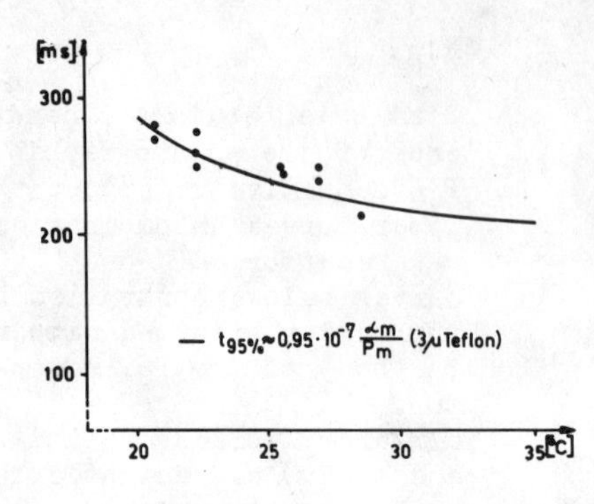

Fig. 3

95 p.c. rising time of a P_{O2} receptor in relation to temperature.
The P_{O2} receptor used has a relative long rising time. The drawn curve has been calculated. The values of α_m (oxygen solubility coefficient in the membrane) and P_m (permeation coefficient of oxygen in the membrane) were taken from statements of SCHULER.

Torr O_2 per ^{o}C. This peak is directed upwards both during rising and during lowering the temperature, indicating that the behaviour of the receptor or that of the thermistor caused by (fig. 2) warming or cooling is somewhat asymmetric.

Another problem involved in compensation of temperature effects is connected with the slope of the response of the P_{O2} receptor used when jumps in the partial pressure of oxygen are produced. An experimental investigation has the following results (fig. 3). The higher the temperature the lower is the rising time of the P_{O2} receptor describing an exponential function [cp. 2].

In the further processing of the P_{O2} signal its rising time is electronically corrected and shortened to 20 p.c. of the initial value by an active network including 3 time constants. It seems to be possible to correct the influences of temperature on the slope of the P_{O2} receptor using only one thermistor placed within the receptor. **For the technical solution of this problem further investigations are necessary.**

Literature

1 KIMMICH, H. P. and KREUZER, F.: Telemetry of respiratory oxygen pressure in man during exercise. in JACOBSON Digest 7 th. Int. Conf. Med. Biol. Engng 1967, p. 90 (Stockholm 1967).

2 SCHULER, R.: Evaluation and design of rapid polarographic in vivo oxygen catheter electrodes. Proefschrift Katholieke Universiteit Nijmegen (Juris, Zürich 1966).

3 WAGNER, W. and SCHNEIDERREIT, R.: Equipment adapted to continuous determination of oxygen consumption in working man. in ALBERT, VOGT, and HEILBIG Digest 10th Internat. Conf. Med. Biol. Engng 1973, Vol. I, p. 141 (Conference Committee, Dresden 1973).

Biotelemetry II. 2nd Int. Symp., Davos 1974, pp. 25–27 (Karger, Basel 1974)

The Dynamic Response of Oxygen Partial Pressure Transducers

Kevin Ewart Forward
Monash University, Clayton

<u>1. Introduction</u>. Oxygen/lead fuel cells operated in a diffusion limited mode can be used as pO_2 transducers, since the output short circuit current of such cells is directly proportional to pO_2. These cells have been described by NEVILLE (1962) and KEY et al (1970). Recently an improved cell has been developed by LAMPARD and FORWARD (1973) which uses an integrated electrode and diaphragm, eliminating the electrolyte layer between them, and giving a faster response time and an immunity from microphonics. In other respects the improved cell and the conventional cell are identical, and both need to be stored in an oxygen free atmosphere or operated into an open circuit, in order to conserve the lead electrode and hence prolong the life of the cell. The following analysis determines the response time of the cell and shows the effect on the output of the cell of storage with an open circuit.

<u>2. Transit Time</u>. The transit time of oxygen through the diaphragm is the primary determinant of the response time when the input pO_2 changes. To determine the transit time, τ_1, consider figure 1. The flow of oxygen, ϕ, is given by the diffusion equation

$$\phi = AD_d \frac{dC}{dx} = AD_d \frac{C_o}{d} \qquad (1)$$

where C is the concentration and D_d is the diffusion coefficient, of oxygen in the diaphragm of thickness d. The velocity oxygen at any position, x, is

$$\frac{dx}{dt} = \frac{\phi}{AC(x)} \qquad (2)$$

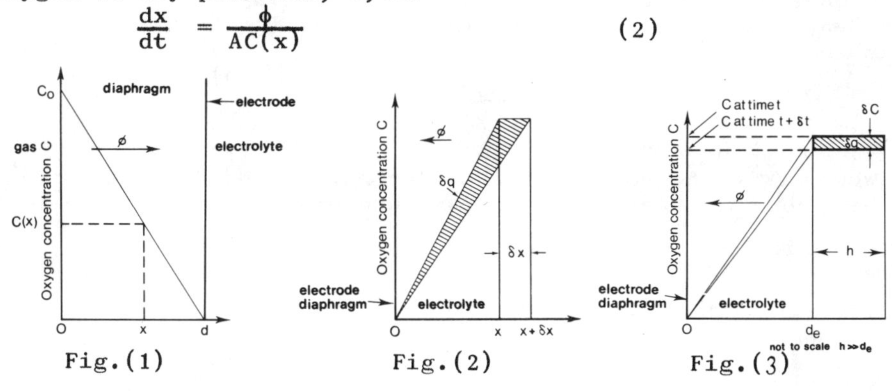

Fig.(1) Fig.(2) Fig.(3)

Solving equations (1) and (2) for the transit time we find that τ_1 is $d^2/2D_d$. Typically this is about 1 second.

<u>3. Removal of oxygen dissolved in the electrolyte</u>. The electrolyte dissolves oxygen, when the cell is not short circuited and is exposed to oxygen, and will deliver current due to the dissolved oxygen when it is brought into operation by applying a short circuit. According to BERGER (1968) the electrolyte will then be uniformly oxygenated except at the electrolyte diaphragm interface where a stagnant layer about 0.5 mm thick exists. To determine the response as a function of time, from the time the cell is first brought into operation consider figure 2, which shows that a quantity of oxygen, dq is removed in time dt by a flow rate, ϕ,

$$\phi = \frac{dq}{dt} \tag{3}$$

and from the diffusion equation (1) we have

$$\phi = \frac{AD_e C_o}{x} \tag{4}$$

It also follows from figure 2 that

$$dq = \tfrac{1}{2}\, AC_o dx \tag{5}$$

Where D_e is the diffusion coefficient of oxygen in the electrolyte, and C_o is the initial concentration of oxygen in the electrolyte. Equations (3),(4) and (5) can be solved to give τ_2 the time taken for the electrolyte layer through which the oxygen is diffusing, to penetrate a distance x from the electrode.

$$\tau_2 = \frac{x^2}{4D_e} \tag{6}$$

$$\phi = \frac{AD_e C_o}{2\sqrt{D_e t}} \tag{7}$$

The time taken for x to equal the thickness of the stagnant layer is about 1 minute, for a typical cell.
When x reaches the edge of the boundary layer the situation then changes to that shown in figure 3. The following equations then apply:

$$\phi = \frac{AD_e C(t)}{d_e} = \frac{dq}{dt} \tag{8}$$

$$dq = VdC \tag{9}$$

Where V is the volume of the electrolyte and d_e is the thickness of the stagnant layer. And hence:-

$$\frac{dC}{dt} = \frac{D_e C(t)}{hd_e} \tag{10}$$

the solution of (11),(12) and (13) is:-

$$C(t) = C_o e^{-t/\tau_3} \tag{11}$$

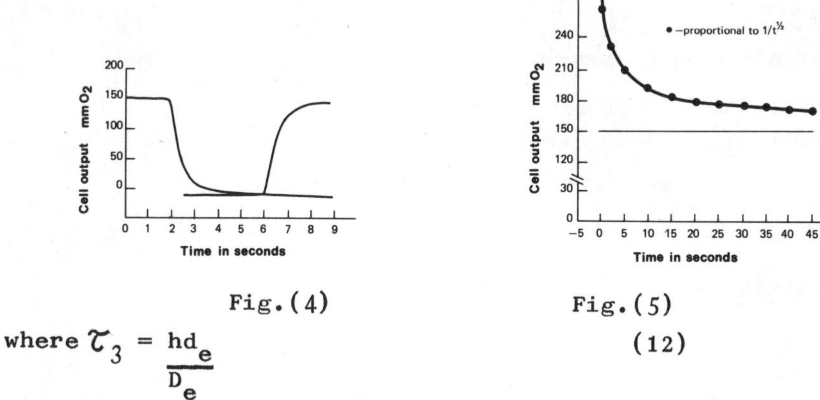

Fig.(4) Fig.(5)

where $\tau_3 = \dfrac{hd_e}{D_e}$ (12)

typically τ_3 is 1 to 3 hours depending on the cell dimensions.

4. Experimental verification. The response of the cell to a step input of 152 mm pO_2 is shown in figure 4. While figure 5 shows the response of the cell to a short circuit after exposure to 154 mm pO_2 while open circuited. Both results are in good agreement with the theoretical responses derived above. Theoretical and measured time constants agree.

5. Implications for operation of the cell. From these results it can be seen that the cell must either be constantly short circuited or short circuited at least 2 hours before it is to be used. The later procedure is to be preferred, as a continuous short circuit considerably reduces the useful life of the cell.

6. References.

NEVILLE, J.R. "Electrochemical device for measuring oxygen", Rev. Sci. Inst., 33, 51, (1962).

KEY A., PARKER, D. and DAVIES, R. "Use of epoxy resin in oxygen electrodes", Phs. Med. Biol., 15, 569, (1970).

LAMPARD, D.G. and FORWARD, K.E. "Fuel cell oxymeters", Prov. pat. appl. (Australia) No. 2997, (1973).

BERGER, C., editor, "Handbook of fuel cell technology", p.138, (Prentice Hall, New Jersey, 1968).

Biotelemetry II. 2nd Int. Symp., Davos 1974, pp. 28–29 (Karger, Basel 1974)

Transducers for Bio-Telemetry

Ved Ram Singh and R. Parshad
National Physical Laboratory, New Delhi

ABSTRACT

Transducers using strain gauges have been described which can measure drip rate, flow, respiration rate and minute body displacements of the order of 2 microns or more. The output of the transducers is in the form of electrical voltage which can be telemetered by the use of the conventional techniques of voltage controlled oscillator, modulation and digital counting.

INTRODUCTION

Some new transducers for measuring drip rate, blood flow rate, respiration rate, and minute displacements have been developed, the output of which,being in the form of a voltage,can be transmitted by conventional telemetry techniques such as use of voltage controlled oscillators, modulation and electronic counting techniques.

In the following is given a brief description of these transducers.

1. Drip Transducer: For giving controlled transfusion of blood, glucose etc., it is necessary to count the number of drops of the fluid going into the patient. The drops of fluid falling on a cantilever obstruction give periodical strains to the strain gauge mounted on the cantilever sheet. The output from the strain gauge is obtained by the usual technique of use of Wheatstone bridge feeding an amplifier-cum-recorder. The output of the amplifier can be fed to a voltage controlled oscillator (VCO) which could modulate a radio-transmitter. On reception, the modulating frequencies are recovered from which, by proper processing, voltage triggers can be obtained which are fed to a digital counter for obtaining the drip counts.

The transducer-cum-telemetry system also serves the purpose of informing the monitoring room if the fluid has been used up during transfusion.

2. Blood Flow Transducer: A strain gauge is mounted on a cantilever mounted in the flow of the fluid whose flow rate is to be measured. The output voltage from the electronic system associated with the gauge would determine the flow rate. For telemetry, the output voltage is fed to a VCO which, in turn, modulates the radio transmitter as in the case of drip transducer. At the receiver, the recovered modulating frequencies are fed to a digital frequency meter, the output of which gives the flow rate in the digital form.

For using the flow transducers in live systems, the vein or the artery, through which the flow rate has to be measured, is by-passed by the flow transducer for the required measurement.

3. Respiration Transducer: The respiration transducer has a steel diaphragm of which one side is attached to the body parts whose respiratory behaviour is to be monitored. On the other side of the diaphragm is mounted a strain gauge connected to an electronic system for monitoring strain.

On application of the transducer to the body, the steel diaphragm takes part in the respiratory motion, which fact gives rise to the corresponding electrical output voltage which, by the technique mentioned above, can be telemetered.

4. Displacement Transducer: The displacement transducer detects small movements of the body parts of the order of 2 microns or more. Thus the transducer can be used to investigate the effect of various stimuli which tend to produce even minute body movements and displacements. The transducer essentially consists of a cantilever sheet on which is mounted a strain gauge. One side of the sheet is fixed and the other free end can be attached to different parts of the body whose movement is to be investigated. As usual, the strain gauge is connected to an electronic system whose output can be telemetered.

For determining the displacement in absolute values (e.g. microns, mm) the output of the electronic system can be calibrated in the following way:

Milligram weights are put on the cantilever and the corresponding output is noted. Next, the cantilever is mounted in an optical system (such as Profile Projector) and milligram weights put on the sheet again on identical sites used for the previous measurement and the corresponding deflection of the free end of the sheet recorded.

CONCLUSION

Transducers for measuring drip rate, flow rate, respiration rate and minute body displacements have been described whose electrical output can be telemetered.

Session II

Microtelemetry – Implants – Integrated Technology

Chairmen: *S. Salmons, R.W. Gill and H. Kunz*

Biotelemetry II. 2nd Int. Symp., Davos 1974, pp. 32–36 (Karger, Basel 1974)

Implantable Integrated Electronics

James D. Meindl
Department of Electrical Engineering, Stanford University, Stanford, Calif.

Abstract

Monolithic integrated circuits present an enormous opportunity for enhanced performance of implantable biomedical telemetry systems. These systems often require micropower operation at low supply voltages, a stringent limitation uncommon in monolithic design; they offer a constant temperature environment within the body, a unique and altogether unexploited asset. Charge coupled devices exhibiting analog memory and time delay promise profound new developments.

Introduction

The almost vanishingly small size of silicon monolithic integrated circuits is their most striking feature. Because of it, these circuits ostensibly are extremely advantageous components for chronically implantable biomedical telemetry systems. However, a cursory study of this possibility reveals that commercial integrated circuits are largely unsuitable for implantable applications because of their large power drain and supply voltage requirements. The principal purpose of this paper is to discuss several strategies for alleviating this problem; these include suitable commercial monolithic circuits, monolithic transistor arrays, monolithic "masterchips", and custom implantable monolithic integrated circuits.

Integrated Electronics in Perspective

Within the past decade integrated electronics--functional assemblies of hundreds and thousands of transistors and related circuit elements fabricated in a minute monolithic silicon crystal--has had a revolutionary impact on electronics. Today, silicon monolithic integrated circuits are ubiquitous in electronics. Our largest computer and communication systems, as well as the smallest electronic wrist watches and pocket calculators, would all be impossible without them. Their paramount importance is due to the fact that they offer markedly improved performance, higher reliability and smaller size and weight all at lower cost compared with alternative methods. The success of integrated circuits can be traced to one principal underlying reason-- their method of fabrication or technology. The many elements of each functional monolithic circuit are produced simultaneously, through a complex sequence of physical and chemical processes, in the exact location in which they are used.

Because of the high initial costs associated with the electronic design, masking and process development for a monolithic circuit, low selling price can be achieved only through high volume production. Consequently, a key goal of integrated circuit design is to achieve a widely useful product which can be manufactured in volume with high yield.

Wide utility generally demands such properties as high gain, wide bandwidth and large dynamic range in linear circuits and large fan-in, large fan-out and high speed in digital circuits as well as a wide operating temperature range in both types of circuits. To meet these demands, particularly while restricting the design to monolithic circuit elements with wide tolerance limits, relatively high power drain and large supply voltages must be accepted.

High manufacturing yield demands that the design of silicon monolithic integrated circuits be absolutely dominated by the requirement that the circuit elements be restricted to small area silicon devices. In bipolar monolithic circuits this effectively limits the available devices to vertical NPN, lateral and substrate PNP transistors, small value ($<$ 10,000 ohms), high tolerance ($\pm$20%) resistors and very small value ($<$ 50 pF) capacitors. In metal-oxide-semiconductor (MOS) integrated circuits virtually all devices are essentially MOS transistors. Consequently, manufacturability requirements severely restrict the available range of circuit elements and hence tend also to increase the power drain and supply voltage of commercial monolithic circuits.

Suitable Commercial Integrated Circuits

At this point, virtually the only commercial monolithic circuits which avoid the preceding problems to a sufficient degree to be useful in a variety of implantable biomedical telemetry systems are the programmable operational amplifier and complementary metal-oxide-semiconductor (CMOS) logic circuits. The programmable operational amplifier is a monolithic circuit whose current drain is controlled or programmed with a single external resistor. Its supply voltage range usually extends as low as $\pm$1.5 v. Circuit performance is, of course, dependent on both current drain and supply voltages. CMOS logic circuits exhibit extremely low standby power drain and a dynamic dissipation which is a linear function of operating frequency. Hence, they offer virtually ideal performance for implantable systems (especially when the CMOS circuits are capable of 1.5 v operation).

The programmable operational amplifier and CMOS logic circuits are now virtually the only commercial monolithic circuits which permit a power performance trade-off which allows a degree of design optimization for an implantable system.

Unfortunately, the majority of such systems require a preponderance of circuit functions which cannot be implemented with these commercial monolithic circuits. Hence, custom-circuit design becomes a necessity.

Monolithic Transistor Arrays

Isolated transistors fabricated in monolithic arrays and separately available at the terminals of a package represent a useful component for implantable systems. They can contribute effectively to small size and high performance in custom circuits for implantable systems when used in conjunction with miniature discrete passive parts. Monolithic transistor arrays have been commercially available for several years and no further discussion of them seems appropriate in this paper.

Monolithic Masterchips

A "masterchip" is a monolithic array of isolated transistors, resistors and capacitors which are uncommitted in terms of a specific metalization pattern for the monolithic elements. Consequently, a custom metalization mask can be designed to achieve a specific interconnection network corresponding to a well defined circuit configuration. Inevitably, some of the monolithic elements in a master chip will be unused in a given circuit; hence, the masterchip is inefficient in terms of functional density. However, it does permit "stockpiling" silicon wafers whose processing schedule is virtually complete. That is, only one custom mask (for metalization) is needed to fully complete the wafer processing schedule. Consequently, turn-around time and cost are small relative to a fully customized monolithic chip for which every mask is peculiar to specific circuit requirements.

On the other hand, compared with transistor arrays, masterchips suffer from the disadvantage that customizing of the chip, in essence, must be done by its manufacturer. This causes inconvenience, time delay and added cost for the user.

Insofar as implantable telemetry systems are concerned, masterchips pose a further problem. The necessity for extremely low power drain requires relatively large values of resistance. Moreover, flexible trade-offs between power and performance demand that a wide range of resistance values be readily available on the chip. Consequently, a "micropower" monolithic masterchip poses special problems of its own. Although its transistors can be employed quite effectively over many decades of operating current, resistance values are much more restricted simply because Ohm's law dictates the current for a specific value of resistance with a fixed supply voltage. The resultant is reduced resistor utilization and functional density.

At the present time, a suitable monolithic masterchip for implantable telemetry systems has not been developed. However, given a micropower masterchip design with extremely high element density and multitap high value resistors, it is quite conceivable that this approach could prove to be widely useful in bringing the advantages of monolithic circuits to a wide variety of implantable systems. Its key advantages are low cost and rapid turn-around time compared with fully customized monolithic circuits. Overall system size may be virtually unaffected by the relatively poor functional density of a masterchip.

Custom Integrated Circuits

A custom integrated circuit is a device whose circuit configuration, mask layout and processing are optimized to achieve a specific set of performance characteristics. It is this class of integrated circuit which brings the full potential of monolithic technology to bear on system requirements. Implantable systems offer a peculiar challenge to monolithic technology which has not yet received sufficient attention. The principal features of this challenge are: 1) The very stringent requirement for low power drain; 2) The desirability of low supply voltages, preferably furnished by a single energy cell; 3) A virtually constant temperature environment; and 4) A highly corrosive environment.

The requirement for low power drain in an integrated circuit implies the use of small geometry monolithic devices--transistors, resistors and capacitors--with very large gain-bandwidth products and extremely low value parasitic elements. Particular attention must be given to monolithic resistors since relatively large values of resistance are required in micropower circuits. Consequently, base or collector pinch resistors, ion implanted resistors or compatible thin film (polycrystalline silicon) resistors are often necessary.

The large impedance levels in micropower circuits offer unusual opportunities for capacitor coupling techniques which are rather foreign to conventional monolithic design. Such opportunities call for compatible thin film capacitors with low parasitics. The newer (e.g. Si gate) MOS LSI (large scale integration) technologies offer promise of monolithic capacitors with significantly larger values per unit area along with low parasitics.

The desirability of low battery voltages to minimize total implant volume imposes added limitations on a monolithic design compared with common practice. Novel circuit configurations and the use of capacitive coupling are necessary to deal with these limitations. The use of an efficient micropower DC-to-DC voltage converter to step-up battery voltages is a worthwhile feature for some systems; others may permit transcutaneous power generation by magnetic induction.

Few environments offer the temperature stability of a deep body location in man or an animal. This opportunity has been virtually unexploited to date in the design of implantable monolithic circuits. The use of simple monolithic RC active networks to simulate inductors is a promising possibility which deserves some further attention. Exceptionally high performance monolithic (tuned) circuits are feasible as a consequence of the controlled temperature environment

The recent advent of charge coupled devices (CCDs) offers an extremely promising opportunity for more sophisticated signal processing in implantable systems. The CCD provides analog memory and time delay which heretofore have been utterly unavailable through monolithic technology. Although no applications of the CCD have yet been made in implantable systems, the prospects for them represent perhaps the most profound opportunity now in view for implantable integrated electronics.

Conclusions

Little has been done to date to exploit the enormous potential of monolithic integrated circuits in implantable telemetry systems. Micropower masterchips and custom low power, low voltage, constant temperature monolithic circuits can add immeasurable capability to implantable systems. The CCD is a totally new monolithic component with profound implications for implantable electronics.

Biotelemetry II. 2nd Int. Symp., Davos 1974, pp. 37–39 (Karger, Basel 1974)

An Integrated Circuit Implantable Pulsed Doppler Ultrasonic Blood Flowmeter

Robert W. Gill and James D. Meindl

Integrated Circuit Laboratory, Stanford University, Stanford, Calif.

One of the principal causes of death in man is cardiovascular disease. There is thus an urgent need to develop instrumentation to assess the performance of the heart and of the circulatory system. While blood pressure is easily measured, the other major parameter, blood flow, is not nearly as accessible; present methods for measuring blood flow are either cumbersome, inaccurate, dangerous or unsuitable for human use. Many applications require chronic experiments with animals; the implantable pulsed Doppler ultrasonic flowmeter described here provides an accurate, non-traumatic technique for flow measurement on a long-term basis.

Fig. 1 shows a simplified block diagram of the flowmeter. A 1μsec burst of 6 MHz sound is transmitted with a repetition rate of 15 kHz through the blood vessel wall and into the bloodstream. A single piezoelectric transducer is used as both transmitter and receiver of the sound. The formed elements in the blood backscatter a fraction of the incident energy to the transducer, with a Doppler frequency shift proportional to their velocities and a time delay proportional to their range. The received signal is amplified, mixed against the transmitted frequency, low-pass filtered and telemetered out of the body. The three one-shot multivibrators ensure phase coherence between the ultrasonic bursts and the timing signal. In the external demodulator, range-gates sample-and-hold the signal at fixed delays after each transmit pulse, recovering the Doppler shifts which correspond to the velocities at different ranges. The instrument thus has great versatility, measuring the spatial velocity distribution across the vessel, mean velocity, vessel size, and volumetric flow, all as functions of time.

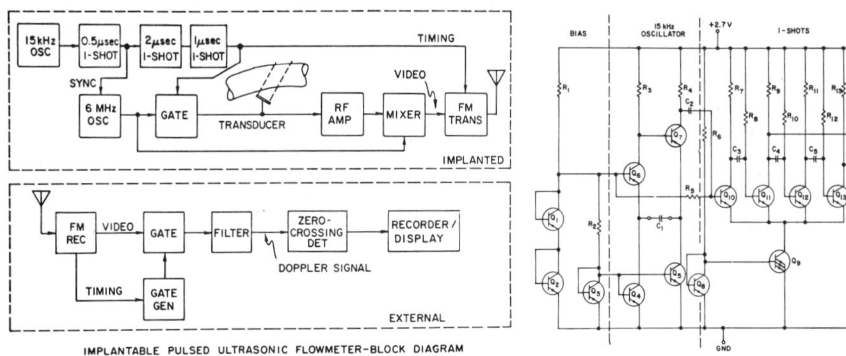

Fig. 1. Flowmeter Block Diagram

Fig. 2. Timer Schematic Diagram

The principal concern of this paper is the implantable portion of the electronics. This package, which must be as compact and reliable as possible, has a considerable number of functional blocks requiring high levels of performance at low power consumption and with a low and variable power supply voltage. These constraints make the use of integrated circuits (IC's) mandatory; by using custom IC's it has been possible to optimize the performance and minimize the physical volume of the package.

Fig. 2 shows a schematic diagram of the timer. By using separate oscillators for the burst-repetition rate and the ultrasonic frequency (rather than achieving phase coherence by deriving the former from the latter with a chain of binary dividers , as is conventionally done), both the amount of circuitry and the power drain are minimized; in addition, this configuration is more flexible than the conventional one. The emitter-coupled oscillator needs only one external timing capacitor (C_1), draws 200μA, and is capable of driving the one-shots directly. The unique one-shot configuration uses only one capacitor per delay unit, plus the input differentiating capacitor C_2; thus it uses a minimum of components and a minimum of current.

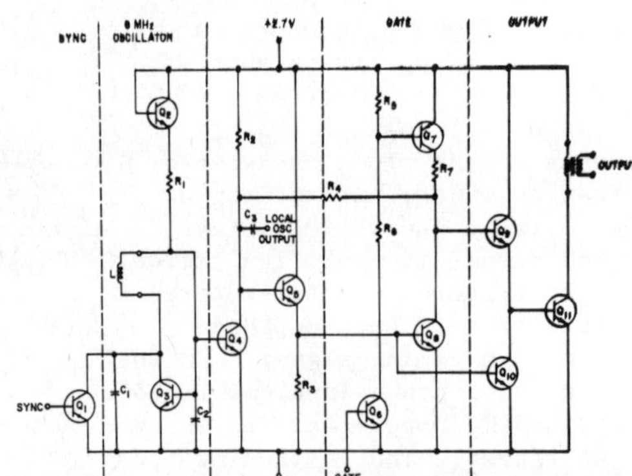

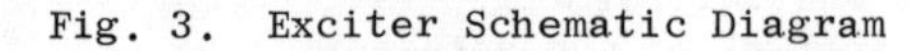

Fig. 3. Exciter Schematic Diagram

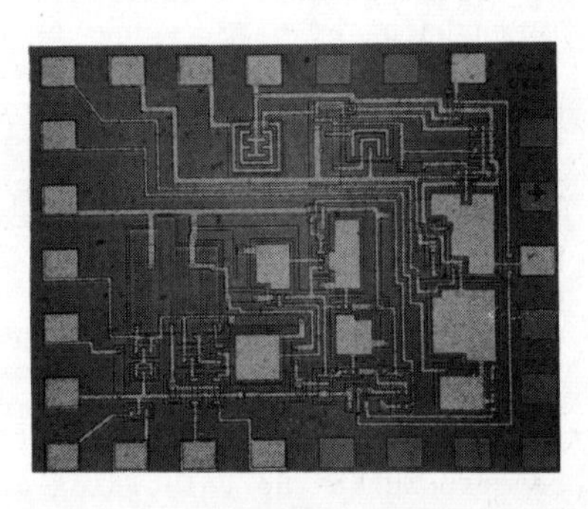

Fig. 4. Timer/Exciter IC Chip.

The exciter schematic diagram is shown in Fig. 3. The oscillator is a modified Colpitts, operating at only 20μA but stable to 0.1% in frequency over a supply range of 2.4 - 3.0 volts. Q_4 is a current-repeater buffer, and Q_2, R_1 and R_2 ensure correct dc biasing of Q_5. The gating scheme, using a switched lateral pnp load for Q_8, ensures that the driver (Q_6-Q_{11}) draws almost no current except during the 1μsec transmit period. The output stage must deliver pulses of current of up to 100mA amplitude, 80 nsec width, with high efficiency, even though the integrated circuit process is optimized for micropower operation. The IC exciter shown in Fig. 4 achieves this drive with 85% efficiency and with great economy of circuit components. Transformer coupling is used both between the driver and the transducer and between the transducer and receiver.

Fig. 5 shows a schematic diagram of the receiver and Fig. 6 the IC receiver; this achieves 74dB conversion voltage gain with 0.8 MHz

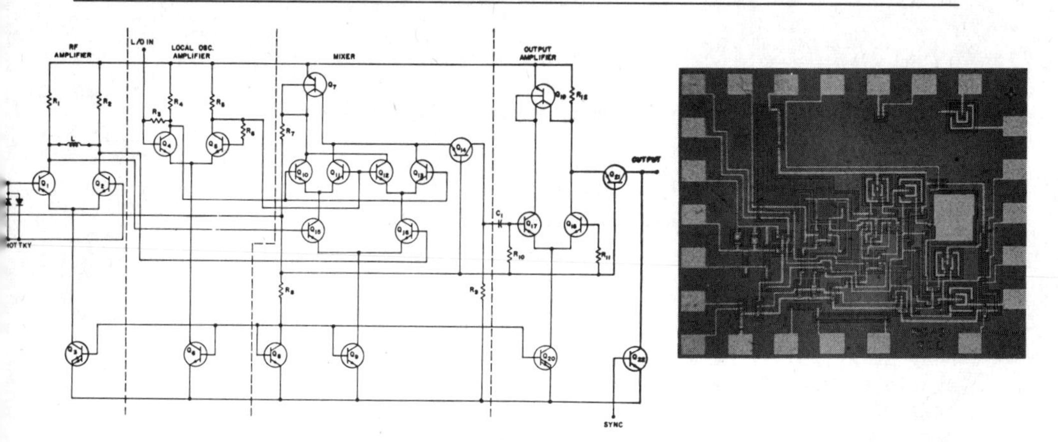

Fig. 5. Receiver Schematic Diagram

Fig. 6. Receiver IC Chip

bandwidth, 2μV sensitivity and 3μsec recovery time, and yet draws only 2.2mW from a 2.7 volt supply. A single tuned rf amplifier stage drives a balanced multiplier mixer, followed by another amplifier stage. Tuning the first stage maximizes its gain, ensures zero dc offset at the mixer input, and minimizes baseband noise originating in the rf amplifier and mixer. The balanced multiplier suppresses the carrier and so minimizes filtering requirements. The bias string Q_7, R_7, R_8, Q_8 not only biases the npn current sources, but also provides the dc current bias for the level-shifter and voltage clamp Q_{14}. Only two external components are required, the input transformer and the tuning inductor for the rf amplifier.

A complete implantable pulsed flowmeter with power supply and telemetry has been assembled using these IC's. Both the power supply and the signal telemetry are achieved using inductive coupling through the skin to a thin flat coil in the implanted package. This technique is probably superior to that using batteries and radio telemetry in the majority of animal experiments and in human patient monitoring applications, since it allows the elimination of both batteries and a custom wideband radio telemetry link. The resulting small package ($28cm^3$) and low power consumption (12mW), in spite of the circuit complexity of ~ 120 circuit elements, indicates the power of IC technology in the development of implantable monitoring instruments.

Acknowledgments

This research was supported in part by PHS Research Grant No. 1P01GM17940-04 from the Department of Health, Education, and Welfare. Among the many people whose assistance has been invaluable the authors would like to mention: H. Allen, P. Curtis, W. Foletta, J. Knutti, R. Taperell.

Biotelemetry II. 2nd Int. Symp., Davos 1974, pp. 40–42 (Karger, Basel 1974)

A Multichannel Implant Telemetry System for Cardiovascular Flow, Pressure and ECG Measurement

Thomas B. Fryer, Harold Sandler and William Freund

NASA, Ames Research Center, Moffett Field, Calif., and Stanford University, Stanford, Calif.

Telemetry systems suitable for implantation totally within the body of an animal have, over the last 15 years, progressed from simple, single transistor devices[1], suitable for the measurement of one physiological variable, to more complex systems[2], capable of measuring many variables using one RF link. Although multiple inputs, as well as signal conditioning, can be handled with relative ease, using recent solid state devices, they have not (to-date) been able to measure an important parameter of the cardiovascular system; namely, flow. The system that is to be described is capable of measuring up to eight parameters, using a high speed, multiplex system (1000 samples per second for each input) and then signal conditioning the analog data to pulse width modulation (PWM) for telemetry via a single RF link. Signal conditioning for one channel of flow, multiple pressure cells, and ECG measurements are included in the system. At present, one channel is sub-commutated for battery voltage and charge current measurements and one for deep body temperature measurement, leaving two spare channels for future expansion.

A complete redesign of the formerly described system[2] was necessary to incorporate the measurement of flow. Since low power operation is essential in a total implant system, the use of ultrasonic transducers would appear attractive. Although extensive use and development of ultrasonic flow techniques have been undertaken in our labs and those of other investigators, further improvements are needed to provide accurate volume flow measurements under all conditions. Consequently, the well-established and proven electromagnetic flowmeter technique was utilized and a low power battery system[3] was devised to achieve the lowest possible power need.

The entire system, shown as a block diagram in Fig. 1, operates from two penlight-size (AA), NiCad, rechargeable type batteries that can be inductively recharged[4]. The multiplexing and signal conditioning electronics are contained within one hermetically sealed case, approximately 2cm x 4cm x 8cm, while the batteries and control electronics[4] are contained in a second hermetic case about the same size, as seen in Fig. 2. A 10cm diameter coil (Fig. 2) encased in medical-grade silastic is used for inductive power pick-up to recharge the battery and to transmit the PWM data out. The entire system is surgically implanted within the thoraxic cavity of an animal. The flow cuff is placed around the ascending aorta, measuring cardiac output minus coronary; one pressure cell is placed in the left ventricle and the other in the ascending aorta. Both dogs (approximately 25Kg)

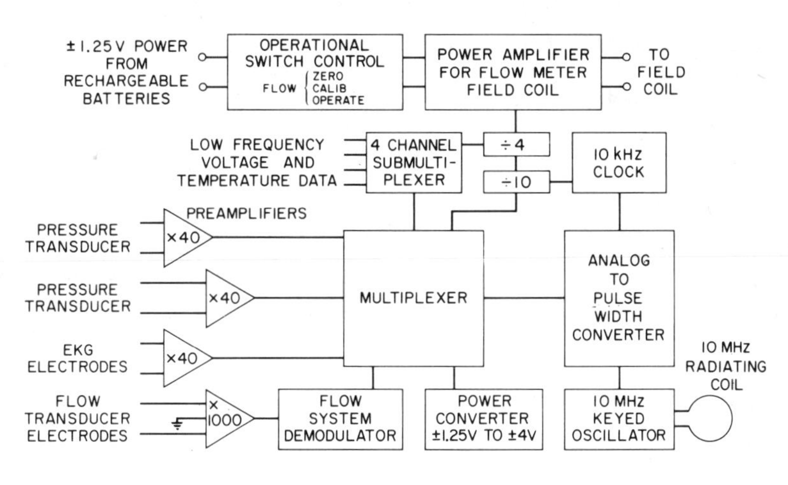

Figure 1 - The Telemetry System Block Diagram

and chimpanzees have been instrumented to-date. The 10cm inductive coil is placed against the rib cage and held in place by the lungs. The recharging coil is located concentrically to the implanted coil and separated a distance of 2-1/2 to 4cm away, depending upon the thickness of the rib cage. The charge coil not only provides for recharging the battery, but also allows turn-on and turn-off control of the system[4] via a frequency discriminator. In the "operate" mode the system can be switched through the intact skin for "flowmeter field on," "field off" and "calibrate." The batteries used have a 500mah rating, so that the 180ma required to operate the system with the flowmeter on provides 2 to 3 hours of continuous operation. But, with the field off (flowmeter zero and not measuring flow), the current drain is only 20ma, so that 24 hours of operation is then possible. By monitoring an experiment using the ECG and pressure, and at a periodic intervals turning the flow system on, operating periods of 6 to 8 hours can easily be achieved. From a completely discharged battery, it takes about 14 hours to provide a complete recharge and this is done with a small, portable, battery-operated oscillator that is attached to the animal via a fitted vest.

The system that has been developed now makes it possible to measure the significant cardiovascular parameters of flow, pressure and ECG from chronic awake, active, unanesthetized animals. Past experience has shown that leads brought through the skin with careful attention are tolerated by some animals, but eventually lead to infection and consequent death. Other animals, such as chimpanzees, will tear at external leads and their use is completely impractical. A total implant system has proven the most reliable for chronic use, but the inaccessability of the required internal power source has prevented its use with transducers like flowmeters that draw large amounts of power. The described multichannel telemetry system can transmit not only ECG and pressure data (low power requirements), but flow data from an electromagnetic flowmeter transducer with its many-magnitudes-greater power requirements. The power capability of this system will allow further expansion of the variety of physiological measurements that can be made in the future, without excessive power limitations.

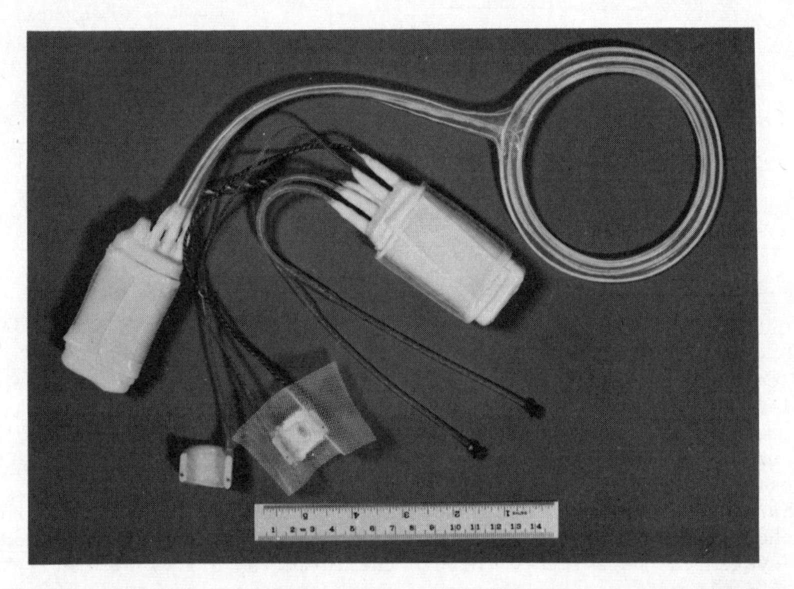

Fig. 2-Multichannel Telemetry System with Flow, Pressure, & EKG Transduders and Rechargeable Power-Pack Prior to Implantation.

REFERENCES

1. Mackay, R. S.: "Telemetering from Within the Body of Animals and Man, Endoradiosondes"; C.A. Caceres (ed.); Biomedical Telemetry; Academic Press, N.Y., Chap. 9, 1965.
2. Fryer, T. B.; Sandler, H.; and Datnow, B.: "A Multichannel Implantable Telemetry System"; Dig. 7th Int. Conf. on Med. Biol. Eng., Almquist and Wiksell, Stockholm, 1967.
3. Fryer, T. B.; and Sandler, H.: "Miniature Battery Operated Electromagnetic Flowmeter"; J. App. Physiol., 31-4:622-628, 1971.
4. Fryer, T. B.; and Sandler, H.: "A Rechargeable Battery System for Implanted Telemetry Systems"; Proc. 23rd An. Conf. Eng. Med. Bio.; p. 129, Washington, D.C. (Nov. 17, 1970).

Biotelemetry II. 2nd Int. Symp., Davos 1974, pp. 43–45 (Karger, Basel 1974)

Telemetry of Radionuclide Tracers by Implantable Thermoluminescent Dosimeters on Rats

Jørgen Bojsen, Ulla Møller, Poul Christensen and Jørgen Lippert

The Finsen Laboratory, Finsen Institute, Copenhagen, and The Danish Atomic Energy Establishment, Risø, Roskilde

The presence of a circadian variation in the 32-P uptake in human mammary tumors is well known and seems to be related to the hormone dependence of the tumor(7). So far conventional measuring techniques have been used for uptake measurements on patients, but recently a two channel radiotelemetrical GM-detector unit for long term surface measurements has been developed (3).

Only few studies of the 32-P uptake(8) have been performed on experimental tumors, usually on rats but kept restrained, which may change the circadian rhythm. These measurements include studies of the 32-P uptake in DMBA induced tumors and in normal skin of rats measured with GM--detectors during 24-hour periods where the well being of the restrained animal was expressed through the persistence of the circadian temperature rhythm measured radiotelemetrically(6).

Measurements of the 32-P uptake performed on unrestrained animals has so far not been reported. Radiotelemetry would be attractive, but it has not been possible for us to reduce the volume of the unit sufficiently to fit a rat(2).

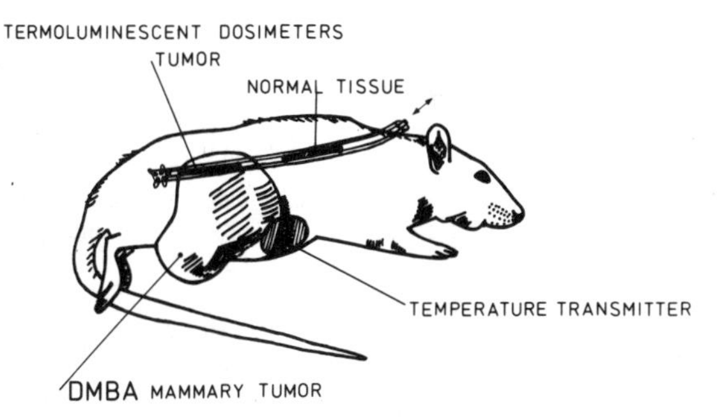

Fig. 1. A sketch of the implanted measuring equipment in a rat with a DMBA breast cancer.

An alternative telemetrical method utilizing thermoluminescence (TL) dosimeters is described in this paper.

These dosimeters can be introduced into the rat through a PVC-catheter and the 32-P concentration can be determined by integration of the dose from the 32-P radiation over adequate time periods. By replacement of the TL-dosimeters, variations in the 32-P concentration can be registrated during a longer period. The dosimeter replacement process may affect the rat and change the circadian temperature rhythm, but this rhythm is continuously measured radiotelemetrically (Fig. 1)(1, 6).

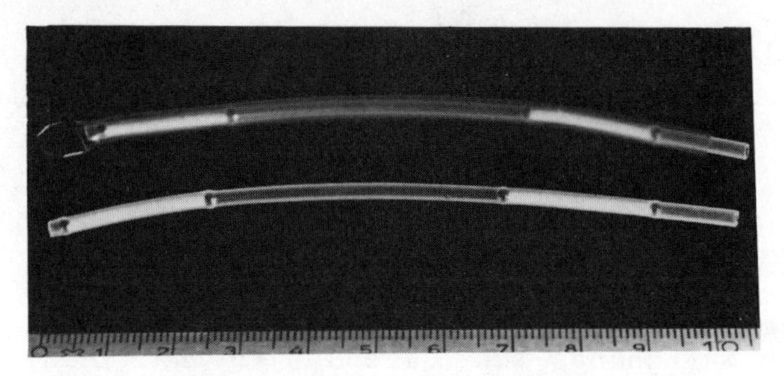

Fig. 2. The double dosimeter unit.

Details on theory and application of TL may be read elsewhere(5). A double dosimeter unit for simultaneous measurements of the 32-P activity in the tumor and in normal skin was prepared by fixing two 60 mg $CaSO_4$:Dy powder samples, a precise distance from each other, in a 2.5 mmØ polyethylene catheter which easily can be introduced through an implanted 3.6 mmØ PVC catheter (Fig.2).The $CaSO_4$:Dy powder was chosen because of its good storage stability and high sensitivity. A laboratory-made read-out equipment(4) was used for reading of the irradiated dosimeters.

Two weeks after implantation of the PVC catheter and the temperature transmitter the measurements can start (Fig.1). The weight of the total measuring system is 5.7 g, 3 % of the body weight of the rat. The thickness of the tissue reaction around the implanted catheter is ranging from 200 to 700 μ. The total thickness of the two catheter walls is 0.8 mm, which permits 32-P measurements. The tissue volumes around the two dosimeters are selectively measured. Several repeated experiments with twelve double determinations based on 1 hour measurements in a polyacrylamide gel phantom with a 32-P concentration of 0.02 μCi/g, corresponding to the concentrations in the experimental animal showed a standard deviation of less than ± 2 %.

An example of the measurements is shown in fig. 3. This demonstrates that the measuring technique is useful for this type of investigation. The method can be adapted and should be useful in other physiological experiments based on radionuclide tracer techniques.

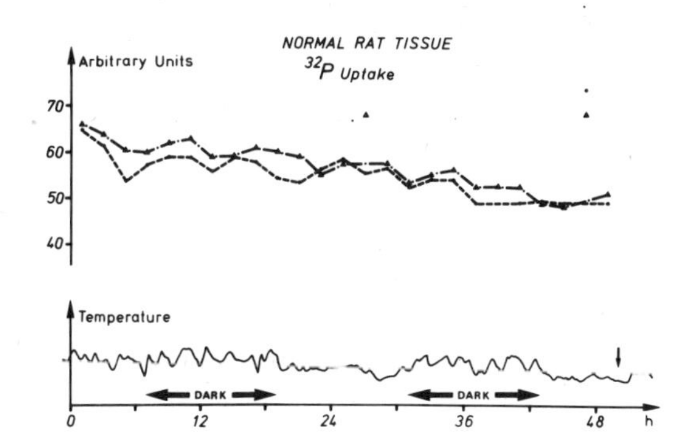

Fig. 3. 32-P uptake measurements during 48 hours with a double TL-dosimeter in normal tissue of a rat, in proportion to the radiotelemetrically measured body temperature

REFERENCES

1)BOJSEN, J., MØLLER, U. and FABER, M.: Radiotelemetrical equipment for continuous subcutaneous measurements of the circadian temperature rhythm in rats.
Pflügers Arch. 328: 176-184 (1971)

2)BOJSEN, J. and WALLEVIK, K.: A radiotelemetrical measuring device, implantable on animals, for long term measurements of radionuclide tracers.
Int. J. Appl. Radiat. Isotop. 23: 505-511 (1972)

3)BOJSEN, J. and VADSTRUP,S.: A portable external two-channel radiotelemetrical GM-detector unit, for measurements of radionuclide tracers in vivo.
Int. J. Appl. Radiat. Isotop. 25: 161-166 (1974)

4)BØTTER-JENSEN, L.: Private communication.

5)MEJDAHL, V.: Risø Rep. No. 249. Proc. III. Int. Conf. Luminescence Dosimetry (Risø, Roskilde, Denmark 1971)

6)MØLLER, U., BOJSEN, J. and LEBEDA, J.: Surface detection of 32-P content of breast cancer.
Proc. III. Int. Conf. Med. Phys., Chalmers Univ. Technol. (Göteborg, Sweden 1972)

7) STOLL, B.A. and BURCH, W.M.: Surface detection of circadian rhythm in 32-P content of cancer of the breast.
Cancer 21: 193-196 (1968)

8)WOLLEY-HART, A., TWENTYMAN, P., CORFIELD, J., JOSLIN, C., MORRISON, R. and FOWLER, J.F.: Changes in 32-P counting-rate in human and animal tumours.
Br.J.Radiol. 41: 440-447 (1968).

Biotelemetry II. 2nd Int. Symp., Davos 1974, pp. 46–48 (Karger, Basel 1974)

Implantable Telemetry System for Long-Term pH Monitoring in the Stomach

P. Baurschmidt, M. Schaldach, J. Greifenstein and F. Stelzner

Dep. für Biomed. Technik, Universität Erlangen,
und Zentrum der Chirurgie der Universitäts-Klinik, Frankfurt/Main

For telemetric monitoring of variations of the pH value in the stomach, endoradiosondes or radio pills have frequently been used. This technique, however, restricts measurements of the acid concentration to only a few hours. To be able to investigate long-term changes in the acid secretion due to alterations of the normal gastric physiology following operations or due to secretory stimulating drugs (1,2,3), it was necessary to develop a small implantable telemetric pH sensor. The device was designed with the goal to keep irritation and disturbances of the normal gastric functioning of the unrestrained animal at a minimum level, while providing at least 300 hours of operating life with the possibility of intermittent transmission for prolonged experiments. In addition, it was necessary to have a linear conversion factor correlating the data signal with the pH value in order to avoid a computerlike and costly interface at the demodulating circuit within the receiver.

Transmitter and Electrode

The transmitter has been designed to operate as an FM-FM system at 102 MHz using a frequency modulated subcarrier for the pH potential/frequency conversion. The potential generated by the pH electrode versus a silver-silverchloride reference electrode may vary between -100 mV at a very high gastric acidity up to -500 mV for almost neutral gastric juice. As pH sensitive electrode, antimony has been chosen because it is difficult to maintain a high isolation resistance over a long time in the case of a miniature glass electrode.

The electrode potential is applied to a DUAL-FET input stage which determines the base current of a subcarrier oscillator. The conversion factor - frequency/pH - is about 200 Hz/pH at a center frequency of 2.2 kHz with a nonlinearity of less than 5% over the full input range. The oscillator is made of discrete parts whereas the frequency modulated RF-stage is a modified FM-Biotransmitter (SANDEV, ESSEX, type SNR 102G) in thick film technique using a Colpitt type oscillator. The current needed for operating up to a range of 50 ft is about 2.5 mA. To power the device with a total transmission time of 300 hours two high efficient mercury cells (MALL. RMCC-1W) and for intermittent operation a magnetical activated switch had to be integrated into the unit.

In order to meet the physiological and technical demands for implanting this telemetric device the complete system had to be split into a transmitter/electrode subunit and a battery part. The connection between both parts is established by use of a 20 cm long and 3 mm wide medical grade silasting tubing shielding two teflon-coated stainless

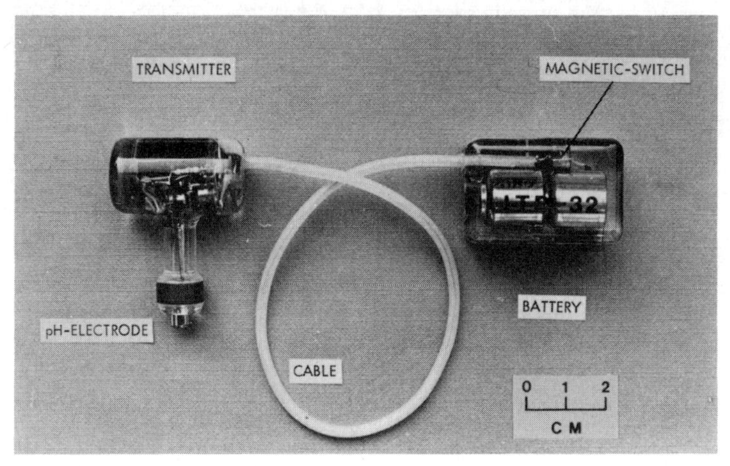

Fig. 1 Transmitter/electrode and battery unit

steel multistrand wires. Both are sealed with epoxy resin (CIBA, Araldit N with Hardener HY 956) which is known from experience with heart pacemakers to cause no tissue reaction and to prevent the electronic components from corroding by body fluid. By splitting the system, a reduction in volume to less than 10 cm^3 and in weight (17 gr) could be achieved for the transmitting part which is sutured on the outside of the stomach with the pH electrode penetrating the stomach wall. Because of the intense muscular activity of the stomach there would be no benefit in further diminishing the size of the transmitter as the fixation in the stomach wall could become a rather difficult task. A photo of the transmitter with the pH electrode and the separate battery unit with the connecting silastic cable is shown in Figure 1.

Receiver

The receiving system utilizes an inexpensive commercial FM-unit with a separate module for demodulating the data signal and for converting it to an analog signal proportional to the measured pH value. Since the transmission bandwith for the pH value is limited from DC to 2 Hz, demodulation is satisfactorly achieved by a pulse averaging circuit. To simplify the experimental procedure when a set of several transmitters is used, the receiver was equipped with an automatic tuning control providing optimal sensitivity for each of the telemetric systems. A separate zero- and gain control is provided to correct differences in electrode potential and to maintain a comparable readout on a strip chart recorder calibrated in pH value ranging from 0 to 10.

In vivo performance

The implantable pH monitoring system described here is been used in a study to elucidate the relationship between an experimentally induced mesenteric vascular occlusion and a resulting change in gastric acidity. In the present study, experiments were carried out with four mongrel

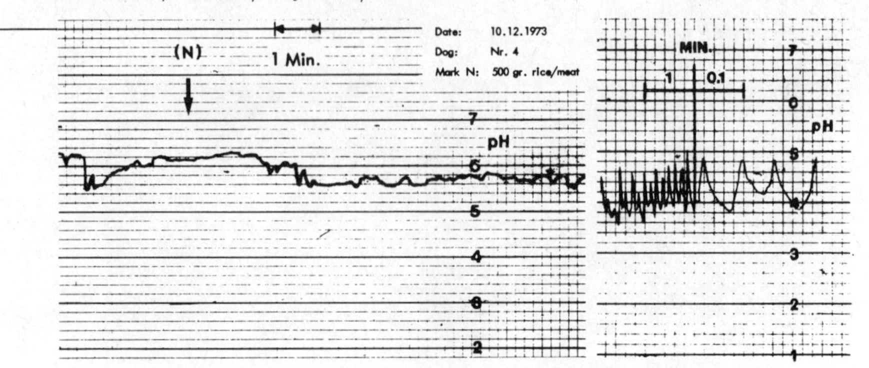

Fig. 2 (left) Recording of pH variation in the stomach of a mongrel dog during feeding
(right) Short-term modulation of pH from muscular activity

dogs, one of them being used as a control. The diminishment of arterial blood flow was regulated by self-occluding constrictors closing the vessel within 50 days. By intermittent operation, the pH signal has been recorded for several weeks transmitting during 1 - 2 hours daily. Figure 2 (left) shows the recording of the pH value two weeks after implantation of the system in a dog weighing 26 kilograms shortly before and after feeding with 500 grams of rice and meat. These measurements of the normal pattern of gastric secretion precede the implantation of an occluder in a second operation and are being used as a control. Figure 2 (right) demonstrates intense modulations of the pH value caused by a muscular contraction rate of 4 - 5 /min. The appearance of this modulation is a result of changes of the positioning of the electrode in the stomach relative to the mucous membrane. It is intended to utilize the experimental technique described here to provide an analysis of this intrinsic pumping rate under different conditions.

Conclusion

The implantable telemetric system for pH monitoring in the stomach allows long-term investigations of changes of the acid secretion due to alterations of the normal gastric physiology, such as those caused by operations or secretory stimulating drugs. The system is capable of operating intermittently with a transmission time of 300 hours up to a range of 50 ft.

References

(1) BOLEY,S.J.;COHEN,M.I.;WINSLOW,P.R.;BECKER,N.H.;TREIBER,W.; McNARAMA,H.;VEITH,F.J.;GLIEDMAN,M.L.: Mesenteric ischemia: A cause of increased gastric blood low, hyperacidity, and acute gastric ulceration.
Surgery 68: 222-230 (1970)

(2) GRÄSER,W.;MENNING,J.;BRUCHNER,H.: A newly developed receiver and procedure of measuring the pH value by telemetry with a endoradiosonde
Biomedizinische Technik 18: 205-208 (1973)

(3) KÜGLER,S.: Langzeitmessung der Magensäure bei Enstehung des hepatogenen Magen-Duodenal-Ulcus.
Habilitationsschrift, Univ. Hamburg (1969)

Biotelemetry II. 2nd Int. Symp., Davos 1974, pp. 49–51 (Karger, Basel 1974)

A Combined pH-Pressure Radiopill

M.A. Ott and U. Faust
Institut für Biomedizinische Technik an der Universität Stuttgart, Stuttgart

For the clinical diagnosis of the stomach and the intestines, the functional parameters of the gastrointestinal tract are very important instruments. These parameters are:

1) acid secretion and pH-value in the stomach
2) neutralisation in the duodenum
3) motility in the antrum and the intestines

To get these values normally catheters are used, which are a considerable physical and psychic inconvenience for the patient which may lead to a falsification of the results. This disadvantage was the reason for developing telemetric methods. Various investigators developed radiopills which make it possible to transmit the pH-value or the pressure from the gastrointestinal tract. For transmitting the pH-value only the "Heidelberger Kapsel", which has an antimony-electrode as transducer, is widely used in hospitals. The pressure has been measured up to now, in a few cases only, by telemetric methods. To get both parameters in one measurement, we developed a combined pH-pressure radiopill, which is able to transmit simultaneously pH and pressure from the gastrointestinal tract.

Glass- and antimony-electrode can be used as pH transducer. The antimony-electrode has the disadvantage of being relatively inexact. Besides, the stability of this electrode depends very much on the temperature. The glass-electrodes are larger and they have a high source-impedance but they are much more exact and stable. If an accuracy

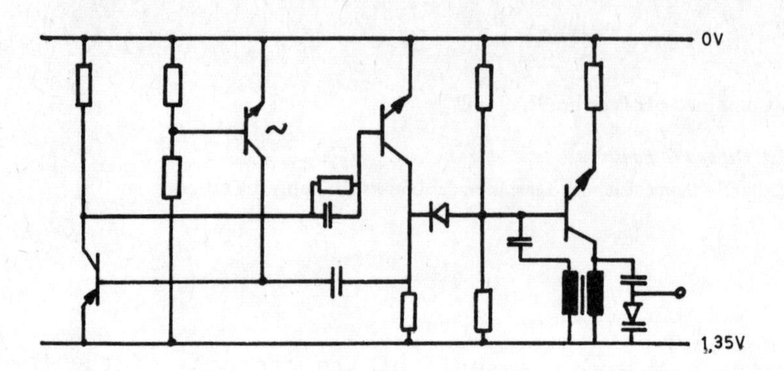

Fig. 1 two channel transmitter

of o.1 pH is demanded glass-electrodes must be used.

Inductive transducers or strain gauges can be used as pressure transducer. We used a pressure sensitive transistor.

With the above transducers a tow channel transmitter was developed. We use a pulse-frequency and a frequency modulation. In figure 1 you can see the complete transmitter. It consists of an astable multivibrator and a HF-oscillator. The voltage of the glass-electrodes modulates the frequency of the HF-transmitter via the varactor . The pressure sensitive transistor controls the frequency of the astable multivibrator, which switches out the transmitter for 100 msec. The multivibrator consists of two complementary transistors and needs only current when the transmitter is switched out. So the battery is equally loaded.

The complete circuit is concentrated on a ceramic disc measuring 8 mm in diameter. Connections between the elements are realised in thin-film technology. Resistors, transistors and capacities are bonded. On the one side of the disc there is the transmitter with the coil, on the other one there is the astable multivibrator.

To avoid negative influences the disc is mantled with a high isolating epoxy. The dimensions of the electronic are 9 mm in diameter and 4 mm in height.

For the radiopill a special receiver had to be developed, because no commercial apparatus was for sale. The receiver must satisfy the following conditions:

1) high sensitivity
2) all-directional reception
3) high selection

These qualities can be realised with a selective receiver. Because the frequency of the transmitter depends on the pH-value, the receiver must follow in its frequency. An automatic circuit can find the frequency of the transmitter and automatically changes the frequency of the receiver. If the input signal of the antenna which is switched on falls below a given threshold, a new antenna with a sufficient input signal is searched. The frequency of the preamplifier and the oscillator is changed by varactors . A ceramic filter causes the selection. After rectifying the IF-signal you get the pulses of the astable multivibrator, which indicate the pressure. A phase detector provides a voltage to follow the frequency. The automatic circuit produces the signal for the antenna selection and the voltage for the varactors . This voltage is amplified and indicates the pH-value.

With this system it is possible to get,in a simple and for the patient comfortable way, pH-value and pressure from the complete gastrointestinal tract.

Biotelemetry II. 2nd Int. Symp., Davos 1974, pp. 52–54 (Karger, Basel 1974)

PDM Multichannel Telemetry System for Biological Use

Tomozo Furukawa, Motoaki Ikeuchi and Goro Matsumoto

Research Institute of Applied Electricity, Hokkaido University, and Mitsumi Electric Co., Ltd., Sapporo

Introduction Demands have been increasing recently for the multichannel telemetry system which enable to measure such biological parameters as the body temperature, the blood flow and the blood pressure as well as the biopotential. Since the absolute values of these parameters are essential, the zero level and the transfer characteristic of the measuring system should be insensitive to variation in ambient temperature and change in supply voltage. This requirement is accomplished in some time division multiplex system by devoting one or more particular channels to the reference purpose: FRYER (1968) and IJSENBRANDT et al. (1971). But this method imposes complex functions upon the receiver system, and beside this, the supply voltage to the reference channels should be stable enough. We developed a PDM/FM multichannel telemetry system which makes the reference channel unnecessary. In the system, the product of frame rate and pulse duration is kept nearly constant regardless of each term being sensitive to the ambient temperature and the supply voltage.

Transmitter The circuit diagram of a 4-channel PDM system is shown in Fig.1. One of channels has dual functions, that is, it is used as a frame synchronization as well. Complementary transistor pairs are used throughout the system to reduce the current drain. In order to make relative variations in the modulator and the timing pulse generator equal, a monostable multivibrator (MMV) and an astable multivibrator (AMV) are adopted. Either of two input units A or B can be attached to the PDM multiplex unit as shown in Fig.1. The unit A is used for the voltage type

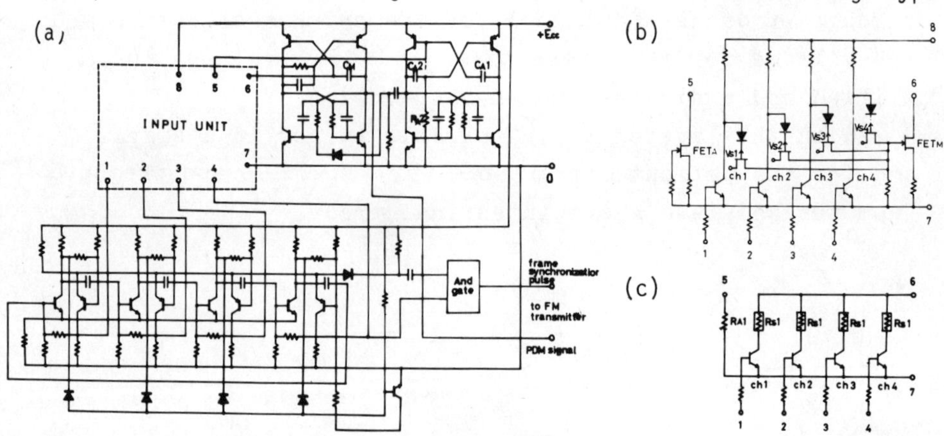

Fig.1 Circuit diagram of 4-channel PDM convertor, (a) main unit, (b) input unit A and (c) input unit B.

transducers and the unit B for the variable resistance transducers. In the case of unit A being used, interval of the timing pulses generated by the AMV is given as follows.

$$T_c = T_1 + T_2$$
$$= C_{a1}E_{cc}/I_{da1} + 2.303R_{a1}C_{a2}\ \log\frac{2E_{cc}-R_{a1}I_{beo}-V_{ces}-V_{bes}}{E_{cc}-R_{a1}I_{beo}-V_{bes}}\ , \qquad (1)$$

where I_{beo} is a base-emitter leakage current, V_{ces} a collector-emitter saturation voltage, V_{bes} a base-emitter saturation voltage and I_{da1} is a discharge current through the constant current source of AMV. Since T_2 is designed in real system to be negligible compared with T_1, $T_c=T_1$ $=C_{a1}E_{cc}/I_{da1}$. Current I_{ds} through the voltage controlled constant current source of MMV is given by

$$I_{ds} = I_{dso} + g_mV_s/(1+g_mR_s) = I_{dso}+GV_s\ . \qquad (2)$$

Then, the modulated pulse duration is expressed as

$$T_s = C_mE_{cc}/I_{ds} = C_mE_{cc}/(I_{dso}+GV_s)\ . \qquad (3)$$

From Eqs. (1) and (3), the demodulated signal is obtained as

$$T_s/nT_c=(C_m/C_{a1})/(K+ GV_s/I_{da1})\ , \qquad (4)$$

where n is the number of channels and K is a ratio of I_{dso} to I_{da1}. In our system, K is designed to be unity. Since the FET used as the current source is operated in the saturation region by a self-biasing resistor, I_{da1} and I_{ds} are unaffected by the supply voltage. The same holds for G if the feedback ratio is large enough. An experimental result is shown in Fig.2.

Next we consider the temperature dependence of demodulated signal. By supposing a temperature dependence of each parameter, we get

$$\frac{T_s}{nT_c} = \frac{C_m exp(a_{cm}-a_{ca1})\Delta T/nC_{a1}}{I_{dso}exp(a_{idso}-a_{ida1})\Delta T/I_{da1} + Gexp(a_g-a_{ida1})\Delta T/I_{da1}}\ . \qquad (5)$$

If $a_{cm}=a_{ca1}$, $a_{idso}=a_{ida1}$ and $a_g=a_{ida1}$ are satisfied, Eq. (5) becomes temperature-independent. The first two conditions are easily satisfied by using capacitors of equal temperature coefficient and a matched pair of FETs. If the feedback ratio is large enough, G becomes near unity, and so, a_g becomes almost zero. In this case, the third condition is satisfied by making a_{ida1} zero. That means the FET should be operated at the drain current where its temperature coefficient is zero. When feedback ratio is not large enough, the condition might be satisfied by making a_{ida} equal to a_g. This can be accomplished by using a source resistor which has a specific temperature coefficient. By using a drain-source resistance of non-biased FET as a part of each biasing resistor, we obtained a variation of 0.04%/°C.

When the unit B is used, quasistable states of both multivibrators are given by

$$T = 2.303RC\ \log((2E_{cc}-RI_{beo}-V_{ces}-V_{bes})/(E_{cc}-RI_{beo}-V_{bes}))\ . \qquad (6)$$

In Eq. (6), the argument of logarithm behaves identically for any transistor of the same characteristics which is operated by the same supply voltage. Consequently, the demodulated signal is given by

$$T_s/nT_c = R_sC_m/\ n(R_{a1}C_{a1}+R_{a2}C_{a2}) = R_sC_m/\ nR_{a1}C_{a1} = K'R_s/\ nR_{a1}. \qquad (7)$$

That is, only a temperature-independent resistor is required to make the demodulated signal voltage-independent and temperature-independent. An experimental result of voltage dependency is shown in Fig.3.

Receiver The frame synchronization is achieved by detecting the modulated pulse which has larger amplitude than the others. A conventional method is used for the demodulation of PDM signals. Since the demodulated signal is inversely proportional to the input voltage V_S as seen from Eq. (4), the original signal is reproduced by an operational circuit composed of a divider and several operational amplifiers.

Discussions The developed system is almost insensitive to change in supply voltage as well as to variation in ambient temperature. This feature prolongs the available battery life or makes it possible to use the rechargeable battery which is not so stable as the mercury cell. Better performanse of the system will be achieved by using the hybrid integrated circuit technology. The transmitter of smaller size and weight, which is essential in biotelemetry, also requires the integrated circuit technology. Consequently, the developed system just meets the general requirements for the biotelemetry. The current drain from batteries can be reduced if the currents through FETs are lowered. But the operating current of FET is so chosen that it gives zero temperature coefficient. This operating current of zero temperature coefficient depends on the type of FET used. In our prototype system, it is 150μA. Overall current consumption without the RF stage is about 800μA for the prototype 4-channel transmitter.

References

FREYER,T.B.: Multichannel implantable telemetry system. NASA Tech Brief, 68-10065 (1968).

IJSENBRANDT, H.J.B.; KIMMICH, H.P.; VAN DEN AKKER, A.J.: Single to seven channel light weight biotelemetry system. Proc. 1st Int. Symp. Biotelemetry, 57-64 (1971).

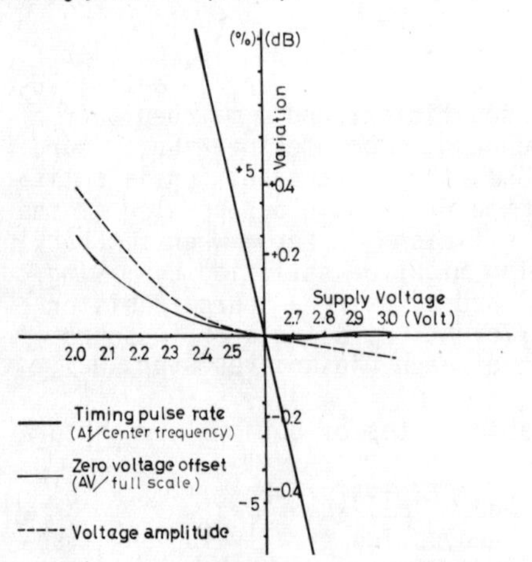

Fig.2 Variation of the different parameters (see insert) vs. supply voltage using unit A.

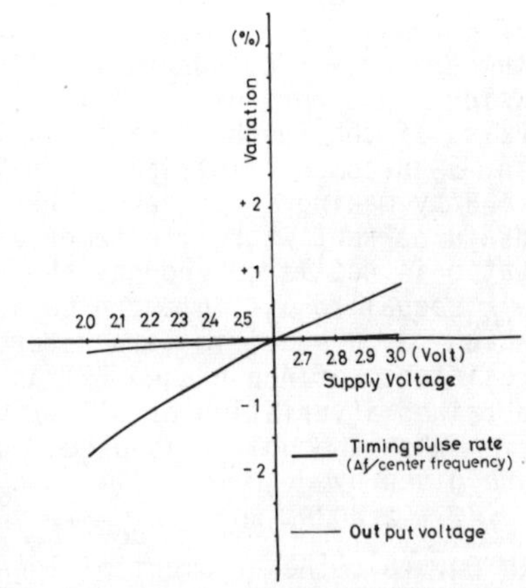

Fig.3 Variation of the different parameters (see insert) vs. supply voltage using unit B.

Biotelemetry II. 2nd Int. Symp., Davos 1974, pp. 55–57 (Karger, Basel 1974)

Low Level 'COS/MOS' Multiplexing for Simplified EEG Telemetry*

R.W. Vreeland and C.L. Yeager
University of California San Francisco Medical Center, San Francisco, Calif.

Conventional techniques require a separate amplifier for each channel in order to obtain a suitable signal level for multiplexing. Elimination of these amplifiers would result in significant savings in both space and in power consumption. Since signals from EEG depth electrodes are several hundred microvolts peak to peak in amplitude, it is possible to multiplex before amplification.

We have tried using single insulated gate FET'S as series input switches. This technique failed because the gating pulses were coupled into the channel via the gate to channel capacitance.

The problem was solved by using CD4016 "COS/MOS" switches. Each switch consists of an "N" channel FET in parallel with a "P" channel FET. The required gating pulses are of opposite polarity. Consequently, the portions of the gating pulses that are coupled into the channel effectively cancel.

Multiplexing at the amplifier input required an amplifier frequency response of DC to 200 KHz for proper sampling. A 312 Hz sampling rate was used. Since a DC amplifier was required, the telemeter can be used for monitoring other low frequency phenomena as well as EEG recording. Actually only three DC channels are provided. The remaining channels have non-polar input blocking capacitors (C1, C2: fig. 1) for removing DC electrode potentials. Similar blocking capacitors are provided for use with the DC channels when DC recording is not required.

The main amplifier consists of two operational amplifiers (A1, A2) operating as followers with gain followed by a differential amplifier (A3). This combination provides a gain of about 120, an input resistance of approximately 200k ohms and a common mode rejection ratio of at least 60 dB. The common mode rejection is adjusted by selecting R3. National Semiconductor NH0001ACF operational amplifiers are used throughout the system. The use of a differential amplifier with good common mode rejection permits bipolar as well as monopolar recording from all nine channels.

The main amplifier (fig. 1) is followed by a non-inverting monopolar amplifier (A4) with selectable gain. The gain and DC bias are determined by R4 and R5 which are mounted on an SM3P Winchester plug. This permits the system gain to be easily changed to accommodate different types of input signals.

*This project was funded by the Office of Naval Research, Contract N00014-69-C-0200-2007.

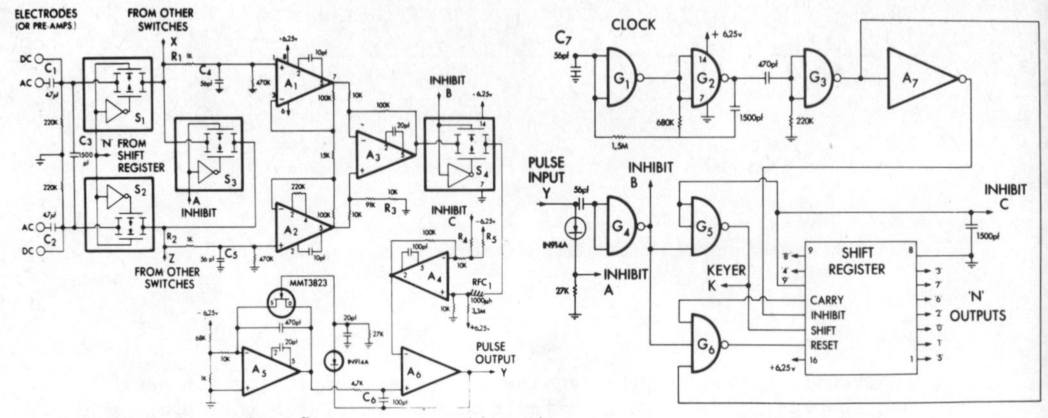

Fig. 1 (left) and fig. 2 (right).

Since differential inputs are provided, two separate switches (S1, S2) are required for each of the nine channels. Only one broad band amplifier (A1,A2,A3,A4) is required for all nine channels. Low power consumption is provided by the low duty cycle pulse position modulator (A5,A6). The transmitted signal is a train of ten pulses followed by a space for synchronization. A sampling rate of 312 pulse trains per second is maintained by the clock (G1,G2).

A somewhat unconventional circuit is used to convert the multiplexed amplitude samples to pulse position modulation. The amplitude samples set the level on the inverting input of a voltage comparator (A6). Its non-inverting input is driven by a ramp generator (A5). The time required for the ramp to run up to the level set on the inverting input of the voltage comparator is directly proportional to the amplitude of the data sample. When the voltage comparator flips, it removes the negative bias from an N-channel Motorola MMT3823 field effect transistor which resets the ramp generator. The duration of the output pulse from the voltage comparator is determined by a 100 pf positive feedback capacitor (C6).

The positive output pulses from the voltage comparator are inverted twice by gates G4 and G5 (fig. 2) before they trigger the CD 4017 shift register. Each pulse shifts the shift register one step thereby enabling a different pair of amplifier input switches.

Zero on the shift register enables channel one, "1" on the shift register enables channel two and so on. The tenth pulse in the train shifts the shift register to position "9" which provides a synchronizing space between pulse trains. The first pulse in the train is generated when the shift register is reset.

The clock (G1,G2) (fig.2) is a CD 4001 connected as a free running multivibrator. It controls the 312Hz sampling rate.

The clock output is differentiated, inverted by G3 and applied to an input to G6. This is a CD 4000 three input NAND gate which generates the reset pulses. Its other two inputs are controlled by the inverted pulses from the voltage comparator and by the "carry" signal from the shift register. All three inputs to G6 must go "low" in order to generate a reset pulse. This prevents premature resetting of the shift register.

Another three input gate (G5) provides the pulses for keying the transmitter and for shifting the shift register. This gate is inhibited during the synchronizing interval by the "9" signal from the shift register. Amplifier A4 is also inhibited during this interval.

Positive pulses from inhibit A turn on switch S3 (fig.1) which short circuits the main amplifier input during the switching intervals. The amplifier output gate S4 is opened by negative going pulses from inhibit B during the switching intervals. This combination of input short circuiting and output gating effectively removes all switching transients.

The transmitter is a Motorola MM4018 transistor operating as a power oscillator in the 88MHz to 108 MHZ FM broadcast band. It operates at one watt peak input power but only sixteen milliwatts average power.

A 2N835 connected as a series switch removes all power from the transmitter between pulses. The six microsecond keying pulses are generated by a CD4001 one-shot (not shown). In order to provide proper bias for the 2N835, this CD4001 is operated from a negative 6.25 volt supply. All of the other "COS/MOS" circuits are powered by a positive 6.25 volt supply. These two power sources are connected in series to provide 12.5 volts for the transmitter. The same sources provide +6.25 volts and -6.25 volts for the operational amplifiers and the scalp electrode pre-amplifiers.

The pre-amplifiers have a gain of fifteen, a common mode rejection ratio of more than 70 dB,an input impedance of one megohm and a noise level of approximately one microvolt RMS. No pre-amplifiers are used for depth electrode recording.

The complete telemeter as used for depth electrode recording draws only 2.2 milliamperes from the 50 milliampere hour nickel cadmium battery. With the FET pre-amplifiers which are used for scalp electrode recording, the total battery drain is only 2.6 milliamperes.

The battery pack consists of ten Burgess CD2 cells potted in epoxy in a 1.7 cm x 3.2 cm x 3.8 cm box. This package weighs 45 grams and has ample capacity for twelve to eighteen hours of continuous operation without recharging.

For 200 to 300 hour battery life, we use two Burgess H146X disposable mercury batteries. Two 1N914A diodes are connected in series with each mercury battery to reduce the voltage to 7.2 volts. Discarded batteries must not be burned; as doing so would release toxic mercury vapor.

Our receiver is an Astro Communications Laboratory SR-209 with a 300 KHz bandpass. A shift register converts the pulse position modulation to pulse width modulation. The nine EEG signals are then recovered by low pass filtering.

The telemeter has been useful for location of seizure foci prior to neurosurgery.

The system performance for depth electrode recording when tested with a 10k ohm source is: Crosstalk 30dB down, noise level 5 to 13 microvolts RMS (determined by noise level in receiving amplifier), maximum input ±one millivolt, frequency response 0.2HZ to 150HZ, input impedance 200K ohms, common mode rejection 60dB. With scalp electrode pre-amplifiers, the performance is: noise level one microvolt RMS, input impedance one megohm, and common mode rejection 70dB.

Biotelemetry II. 2nd Int. Symp., Davos 1974, pp. 58–60 (Karger, Basel 1974)

An Electronic Instrument for Gastro-Intestinal Telestimulation

Károly Bretz
Research Institute of Physical Education, Budapest

An electronic device was developed which consisted of an orally introduced miniature capsule for telestimulation purposes, control circuits, "primary" coils, EEG transmitter, receiver and recorder. Evoked brain potentials have been recorded on restrained and unrestrained cats when affected by the electrical stimulation of the small intestine.

Introduction

Already in the beginning of the 1930's, several researchers had dealt with the realisation of telestimulation. The possibility of solution was seen in the realisation of the electromagnetic induction procedure. The primary coil was placed near the tissue desired to be stimulated and the secondary coil below the skin in an operative way. LOUCKS /1933/, CHAFFEE and LIGHT /1934/ have been pioneers of the technique. A radiofrequency inductive energy transmission was employed by GLENN et al. /1959/. A new procedure was elaborated by DELGADO /1963/ for the purpose of telestimulation of the brain structures. Concerning the correlation of cortical stimulation and the visual cortex excitation BRINDLEY and LEWIN /1968/ reported on interesting experiments.

A new technique was introduced by WECHSLER et al. /1967/. The characteristics of the radiofrequency inductive transmission through the skin and the quantitative data of the losses were discussed by SCHUDER et al. /1971/.

A wired system was employed by BILGUTAY et al. /1963/ and KATONA /1963/ for the gastro-intestinal pacing to introduce a new concept in the treatment of paralytic ileus by direct electrical

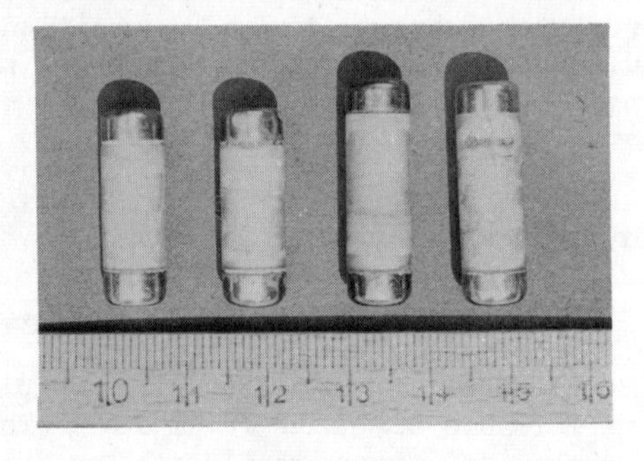

Fig. 1.
Ingestible capsules for gastro-intestinal pacing

stimulation. The topography of the interoceptive integration was investigated by ÁDÁM /1967/. The up-to-date current-supply methods have been discussed by FRYER /1974/.

In our own experiments we have developed stimulating capsules. We have employed these simultaneously with the wired stimulating installation, or independently - in some cases - on restrained and unrestrained cats. We have stimulated the small intestine with single impulses or impulse-series and have examined the evoked brain potentials.

Material and methods

Twenty-five adult cats were employed. Twenty two cats were anesthetised with chloralose and the electrocortical activity induced by the small intestine stimulation was directly led from the surface through silver macro-electrodes. The other experimental animals were unrestrained and equipped with chronically implanted electrodes.

We have utilized DISA Multistim generator, connecting circuits, monostable multivibrators, amplifiers, primary coils, capsules, EEG transmitter, receiver and recorder. The pulse series produced in the primary coils /duration: 1 msec, peak amplitude 5 A, 40 pulses, series time: 120 sec,/ induced the stimulating voltage in the capsule in an inductive way /Fig. 1., 2./. In the capsule a voltage limiting and rectifying mean may be set.
In order to minimize the effect of the capsule movements, we employed three primary coils which were excited by the monostable multivibrator - amplifier system practically without any time delay. The two stimulating electrodes on the capsule were gilded and produced 0 - 5 V values, direct electrical stimulation.

The evoked brain potentials we have, with a one-channel telemeter, transmitted into the receptor-device to which an amplifier and Intertechnique multichannel analyser are joined.

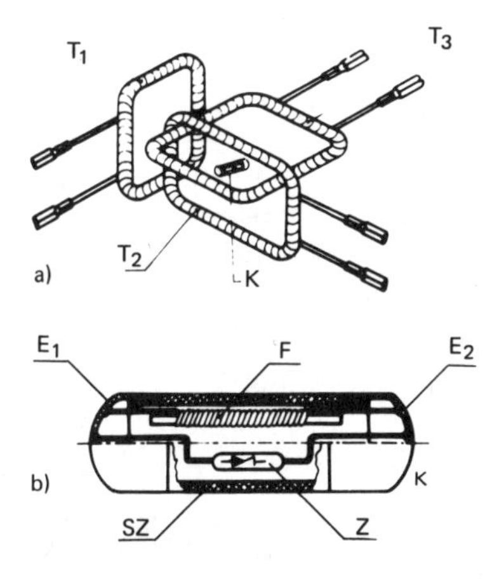

Fig. 2.

Layout of the coils, capsule and circuits

a/ Coil arrangement
b/ Capsule

$E_{1,2}$ gilded electrodes
F coil
SZ plexiglass tube
Z Zener diode

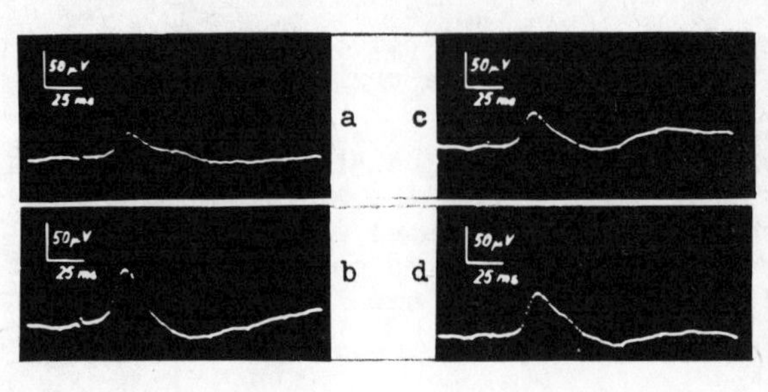

Fig. 3.
Evoked brain potentials recorded by wired pacing system and telestimulation induced in the small intestine
a, b, : wired stim.
c, d, : telestim.
U_{st} : 1,5 V and 2 V

Results

We have dosaged impulses with 3 sec pauses, forty times. Their amplitude was max. 5 V. The afferent electrical activity induced with the stimulation of the small intestine arrived into the higher brain centres. The punctum maximum could be found on the somato S II area. The maximum value of the evoked brain potentials was 100 uV. Threshold value of the stimulation amplitude is about 1 V. The latency times were functions of the amplitude of stimulation. It became evident that the form of the stimulating pulses were influenced by the electrode polarisation. The changes of pulse-forms were, however, not reflected on the form of the evoked potential average. These potentials were influenced by the peak-value and integral of the stimulation pulses. The evoked potentials wined with a wired system and with telestimulation have shown a good correlation / r: 0,85 - 0,95 /.

The author is grateful to Prof. Dr. G. Ádám for his ideas and support during the course of this investigation /Eötvös Loránd University, Budapest/.

References

ÁDÁM, G.: Interoception and behaviour. Publ. House of the Hung. Acad. of Sci., Budapest (1967).
BILGUTAY, A.M.; WINGROVE, R.; GRIFFEN, W.O.; BONNABEAU, R.C., and LILLEHEI, C.W.: Annals of Surgery 158: 338-348 (1963).
BRINDLEY, G.S.; LEWIN, W.S.: J. Physiol. 194: 54-55 (1968).
CHAFFEE, E.L., and LIGHT, R.V.: Yale J. Biol. Med. 7: 83-128 (1934).
DELGADO, J.: in SLATER, L. /ed./: Bio-Telemetry, 231-25o (Pergamon P., New York 1963)
FRYER, T.B.: Biotelemetry 1: 31-40 (1974)
GLENN, W.W.L.; MAURO, A.; LONGO, E.; LAVIETES, P.H., and MACKAY, F.J.: J. Med. 261: 948 (1959).
KATONA, F.: Thesis. Hung. Acad. of Sci., Budapest (1963).
LOUCKS, R.B.: J. Comp. Psychol. 16: 439 (1933)
SCHUDER, J.C.; GOLD, J.H., and STEPHENSON, H.E.: IEEE. Trans. BME 18: 256-273 (1971)
WECHSLER, R.L., and GARBER, H.J.: Med. Res. Eng. 1st Quart. 34 (1967)

Session III

Storage Telemetry

Chairmen: *A. Shah and F. Pellandini*

Biotelemetry II. 2nd Int. Symp., Davos 1974, pp. 62–63 (Karger, Basel 1974)

Storage Telemetry: Session Introduction

Arvind Shah
Institut für Technische Physik, ETH Zürich, Zürich

Defining Storage Biotelemetry as the "acquisition of biomedical parameters with the help of recorders carried by the sportsman, patient or animal under observation", we may point out that until now the application of the technique has been limited to cases where

a. the distances between patient or test animal and the observer are so great or unpredictable as to render difficult the reliable use of wire and even radio telemetry: This will often be the case with free-ranging animals which might migrate over hundred of kilometers or with ambulant patients (who may, for example be living their normal working life and commute within a large city)

b. the medium in which the sportsman, patient or test animal is situated is such as to exclude radio telemetry: This is the case with the underwater situation; as is well known, radio waves are hardly transmitted in water, and sound waves are practically the only way of underwater communication, the corresponding technology being rather cumbersome.

As for the methods of recording used, these again can be divided into 2 classes:

1. Analog Recording Instruments,
   meaning normally electromechanical recorders using magnetic tape as a recording medium

2. Digital Recording Instruments,
   meaning fully electronic equipment using Digital Integrated Circuits as a recording medium

Both instrumentation categories listed above have made substantial technological progress in the last years: On one hand, subminiature electromechanical tape recorders (e.g. Avionics, Marquette, Oxford) have been developed and on the other hand low-power integrated memories (complementary MOS memories) are becoming available with higher capacities (up to 1 kBits per chip at present). These developments are rapidly making storage telemetry more attractive and opening new possibilities for applications.

Reviewing the four papers of the present session, one can see that Pilani et al's contribution (III/1) falls into the category of underwater telemetry (application b) and uses the tape recorder (method 1); Mc Kinnon (III/2) describes the subminiature tape recorder developed at Oxford (method 1); Halm and Yee (III/3) deal with a digital medication chronolog and Pinösch et al (III/4) with digital recording of body activity on free-ranging deer (both application a and method 2). These four papers may therefore be considered a small but representative cross-section of the whole field. The techniques and solutions described therein might thus become useful hints for other biotelemetric applications where one would like to substitute a cumbersome wire link or an unreliable or otherwise difficult radio-channel with on the spot recording of the data.

Biotelemetry II. 2nd Int. Symp., Davos 1974, pp. 64–66 (Karger, Basel 1974)

Instrumentation for Underwater Data Acquisition

Andrew A. Pilmanis, Roland D. Rader, John K.C. Henriksen and John P. Meehan
University of Southern California, School of Medicine, Department of Physiology, Los Angeles, Calif.

In the past, experimental data on the physiological changes in man exposed to an underwater environment have been obtained either by simulating underwater conditions in a hyperbaric chamber or by acquisition of data through the use of "hard wire" methods. The system described in this paper consists of a subject-carried miniaturized tape unit that has been converted to an 11-channel FM tape recorder and further adapted for underwater use. Each channel consists of a signal conditioner and a low frequency oscillator. The signal conditioner conditions the signal and causes the oscillator to be modulated in frequency proportional to the level of the primary signal. The variable frequency from each oscillator is summed and the composite is recorded on the tape recorder.

The tape unit is a Sony BM-10 with an upper limit of frequency response of 10 KHz. The tape recorder and electronics are shown in the open package in Figure 1. The case is an aluminum shell with a quick-release lid and terminations for signal leads. This equipment is completely self-contained and allows complete freedom of movement by a subject, either in the free-diving state or with SCUBA equipment. Any combination of 11 physiological parameters for which there are available sensors may be recorded. Such parameters may include EKG, EEG, EOG, EMG, pressure, temperature, and heat flow. The miniaturized input amplifiers are interchangeable and have a variable gain control, giving the system good versatility. This unit has six connector inputs for sensor attachment, an external stop-start control lever, and an external event-marker switch. It is built to withstand depths in excess of 200 ft; weighs 10 lb dry, 4.5 lb immersed; measures 8 x 17 x 24 cm, and is attached to a SCUBA tank. A voltage analog of the primary signal is obtained by playing the tape into a bank of discriminators and associated tape speed-compensating amplifiers. The outputs are then recorded on a strip chart (Fig. 2).

Standard IRIG frequency subcarrier oscillators are used. These units have been designed to achieve a frequency of oscillation that is independent of temperature, supply voltage, and load changes. They also have inherent frequency modulation limits for the prevention of overmodulation into adjacent channels.

The bridge signal conditioner is a pulse amplifier with a differential input and an excitation pulse pulse generator. It is used to

Fig. 1. Pictorial view of recording device without the lid. An input cable and an ambient pressure sensor are at the left. The electronics are at the top and the start-stop control, at the bottom left.

measure all temperatures and pressures. The excitation pulses are synchronized to the output of their respective subcarrier oscillators. The pulse width is fixed, causing the duty cycle to vary from 0.1% to about 1.5% as the frequency of the subcarrier goes from 400 Hz to 7350 Hz. When a 500Ω bridge is used, the current drain at 6.5 V is approximately 500 μa.

The biopotential amplifier is an AC-coupled differential amplifier. The gain is modest as the amplifier normally functions only to condition an electrocardiographic signal. Total current drain at 6.5 V is 150 μa. If required, the gain and bandpass can be altered before fabrication to permit detection of EMG or EEG.

In the field of experimental diving physiology, many research situations do not require real time underwater monitoring of physiological parameters. In these cases, the underwater recording system

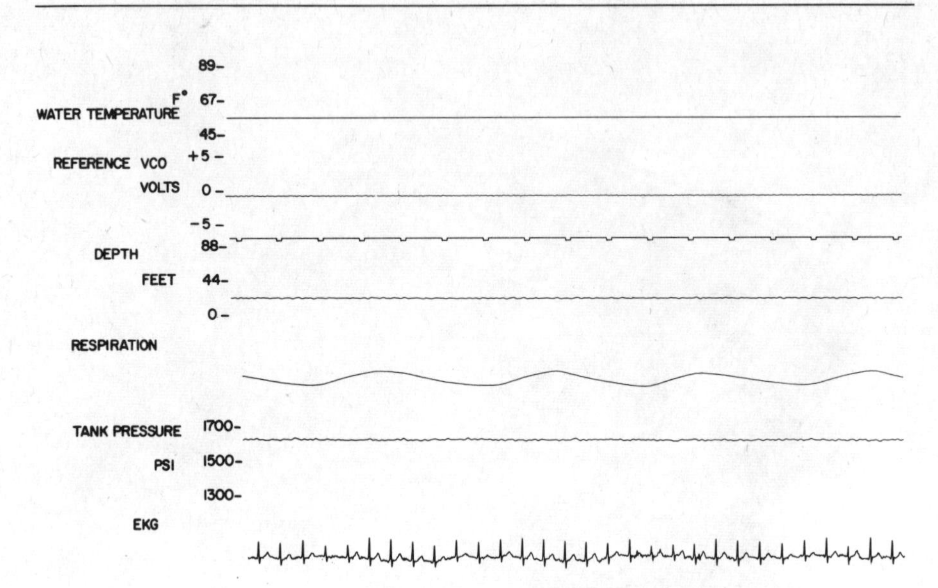

Fig. 2 From top to bottom, strip chart recordings of water temperature, reference VCO output voltage, diver's depth, diver's respiratory signal, tank pressure, and diver's EKG.

has proven a practical method of data acquisition. Unlike underwater telemetry, this system is not hampered by loss of data due to loss of transmission and is therefore able to consistently record good quality data without limiting its boundary. Physical data, such as water temperature and pressure, and physiological parameters can be simultaneously and continuously recorded on tape during the dive and recovered and analyzed after the dive (Fig. 2). This unit is currently being used in investigations of the physiological aspects of underwater exercise and in the evaluation of factors associated with the formation of vascular bubbles during ocean diving.

This research was supported by the Janss Foundation and in part by the Office of Naval Research Contract N00014-67-A0269-0026, with funds provided by the Naval Bureau of Medicine and Surgery, Los Angeles, California.

Biotelemetry II. 2nd Int. Symp., Davos 1974, pp. 67–70 (Karger, Basel 1974)

A Miniature 4-Channel Cassette Recorder for Physiological and Other Variables

J.B. McKinnon

Oxford Instruments, Oxford

Introduction

The use of electronic instrumentation for monitoring patients in hospitals is a well established technique. A vast range of monitoring and recording equipment is available to collect, record and display physiological data on patients who are immobilised and in bed or restricted to the confines of the ward.

In the particular case of cardiac monitoring, in many cases patients who have intermittent abnormalities which are not of a serious nature are immobilised unnecessarily in order that "phantom" arrhythmias or evanescent electrocardiographic changes can be observed and identified.

In this case, the use of a portable long-term recording system not only eases the demand on the Coronary Care Unit but also allows the patient to continue his normal activities and in many cases to go home whilst monitoring continued. The system described allows complete freedom for the subject and provides continuous four channel recording for periods of 24 hours or more.

Design Considerations

A number of considerations influences the design of the system from the outset :

1. Continuous recording capability for 24 hours or more.
2. Rugged compact construction.
3. Reasonable data accuracy.
4. High speed data replay.
5. Simple operation
6. System flexibility to allow a wide range of parameters to be monitored.
7. The use of readily available commercial tape cassettes.

All the above features were incorporated into a pocket sized unit which can be carried by the patient with very little inconvenience. Operation of the recorder is simple and tape cassettes and batteries are easily changed by the subject with the minimum of instruction to allow surveillance to be maintained for many days if needed.

The Recorder : The basis of the system is a miniature four track recorder which operates for 24 hours continuously using standard C120 cassettes. This recorder is the result of a joint development effort between Oxford Instruments and Dr. F.D. Stott of the Medical Research Council Clinical Research Centre, Northwick Park. The construction of the recorder is shown in Fig. 1.

A miniature D.C. motor, the speed of which is electronically controlled, drives the capstan directly through a gearbox and the tape take-up via a slipping clutch. Power is supplied by mercury cells which give sufficient power reserve so that some types of transducer can be energised directly from the recorder. The recorder also accomodates four separate modular plug-in amplifiers all within the dimensions of 110 x 86 x 36 mm and a total weight of 400 g.

Two types of recording amplifier are used to cope with the wide range of possible input signals, one using the direct recording method familiar to users of audio equipment, and the second utilising a pulse width ratiometric technique to give a D.C. response.

Direct Recording : Parameters such as ECG, EOG and EEG, needing a greater system frequency response than that possible with the carrier method but no particular absolute amplitude accuracy requirement, are recorded using the direct recording method. Using this approach, the main technical problems occur at the signal recovery stage due to the fall off in replay head sensitivity with decreasing frequency. For this reason, in normal audio recording the low frequency limit of a tape system is around 30 - 50 Hz, and even with a specially designed cassette head the replay lower frequency limit (3 dB) of the Medilog system is around 10 Hz (See Fig. 2). In order to recover the low component frequencies of ECG etc. which are actually recorded on the tape, the cassettes are replaced in accelerated time. This procedure has two advantages : (a) long-term information is recovered rapidly and (b) all recorded frequencies are multiplied by the record-play ratio (typically X 60) and can be detected by the replay head.

Fig. 3 shows the typical low frequency performance which can be achieved by this means. In fact, it is generally more useful to deliberately restrict the low frequency response of the record amplifiers to eliminate interfering baseline shifts to subject movement.

Carrier Recording: A Pulse width ratiometric PDM technique is used in order to achieve reasonable level of signal amplitude accuracy without the need for tape flutter compensation channels. Since the capstan is directly gear driven and the tape recorder is often used in conditions of relatively high vibration and shock, short term tape speed variations can be a few per cent. The P.D.M. method inherently rejects the effects of tape speed variations provided the flutter frequencies are not close to the carrier frequency. Fig. 4 shows the principle of the method. Using this technique, four channel recording can be achieved with S/N ratios in the range of 30-40 dB.
One special function of the basic P.D.M. amplifier is used to record

resistance changes directly, so that pressure transducers and thermistor temperature sensors may be plugged directly into the recorder inputs without additional signal conditioning amplifiers, while another version may be used to generate a stable clock channel. This clock module has a dual function. Besides providing absolute timing information independent of record and replay tape speed settings, the unit is also used to generate event marks in the time channel by external push-button actuation.

Data Retrieval : The miniature tape loggers are record only devices, so that a separate laboratory instrument must be used to replay the taped information in accelerated time. The replay is of modular construction so that the configuration of replay amplifiers can be arranged to suit that of the recorders. For example, a typical configuration might be :

- Channel 1 - ECG (direct)
- Channel 2 - EEG (direct with external preamp.)
- Channel 3 - Blood Pressure (carrier)
- Channel 4 - Time and Event (carrier)

Early users of the system analysed their replayed data by visual scanning on oscilloscopes and high speed write-out using ultra-violet recorders to produce hard copy. (Ref. 1 & Ref.2). However, since each cassette contains the equivalent of 2 kilometers of 4-channel chart paper, obviously automatic methods of reviewing the data and condensing this into readily assimilable forms are required.

In the particular sphere of cardiology an automatic ECG analysis system for processing 24-hour electrocardiograms has been developed. This analysis system is the subject of a recent paper (Ref. 3). More recent work includes expansion of the analysis unit to multi-channel capability, particularly with a view to the analysis of simultaneous records of ECG and direct inter-arterial blood pressure waveforms.

Summary : This short paper has attempted to describe the capabilities of a long-term tape recording system. Actual and potential applications are so wide that only a few can be mentioned. The system is already in use recording such parameters as temperature, vibration, noise level, depth underwater and geomagnetic phenomena as well as physiological parameters. With such a system it is possible to cross-correlate multichannel data to obtain a picture of physiological responses to real life situations. The main problem remaining is to match the information gathering capacity of the system with equally powerful methods of data analysis.

LITTLER et al "Continuous Recording of Direct Arterial Pressure & Electrocardiogram in Unrestricted Man", B.M.J. 3. 76-78 (1972).

LITTLER et al "Direct Arterial Pressure & Electrocardiogram in Unrestricted Patients with Angina Pectoris", Circulation XLVIII:125-134(1973).

CASHMAN P.M.M. and STOTT F.D.A., "A semi-automatic system for the analysis of 24 hour ECG recordings from ambulant subjects", Biomedical Engineering 8: 54-57 (1974).

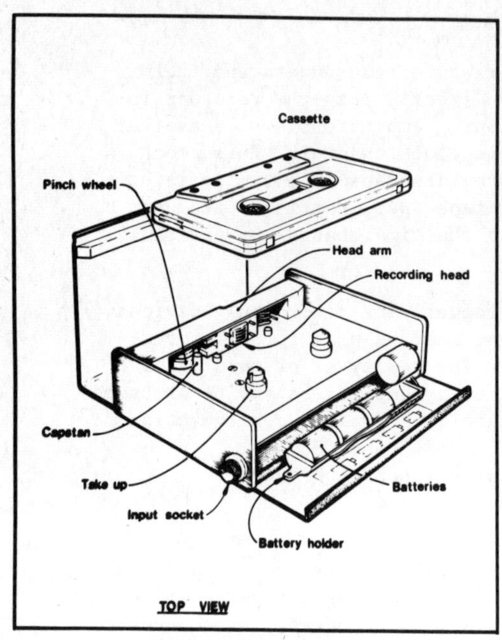

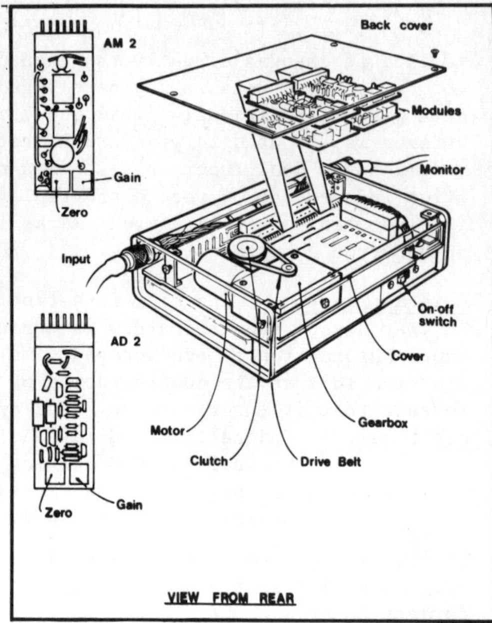

FIG 1 The MEDILOG 4-Channel Recorder

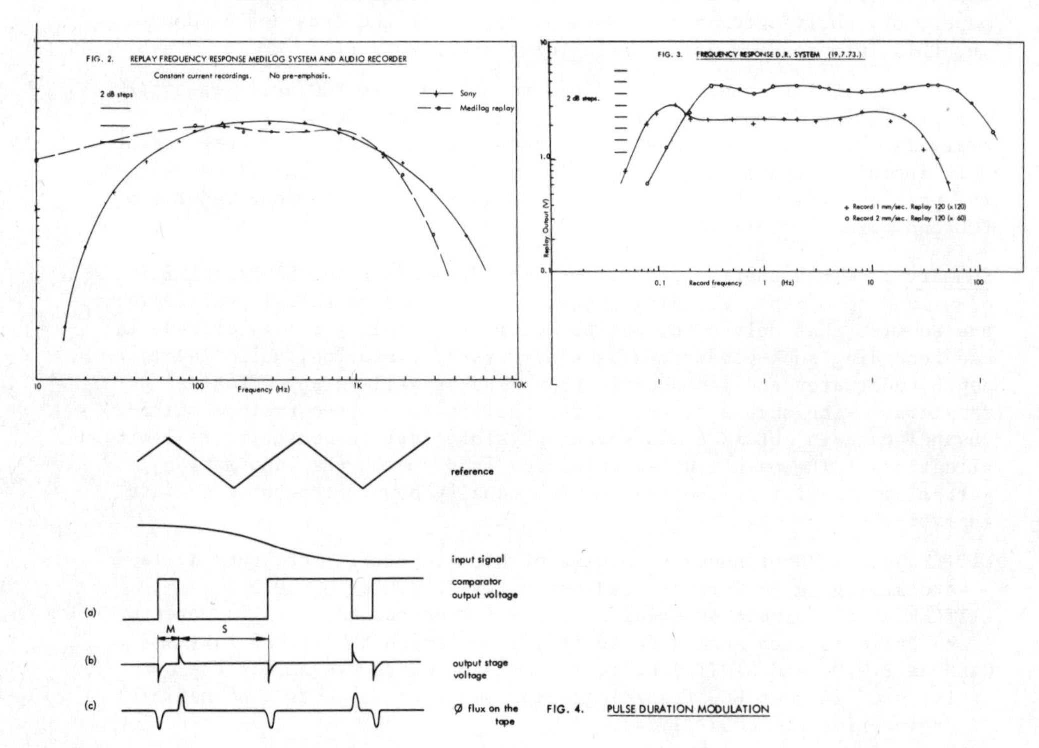

FIG. 4. PULSE DURATION MODULATION

Biotelemetry II. 2nd Int. Symp., Davos 1974, pp. 71–73 (Karger, Basel 1974)

Medication Chronolog

Pierre M. Hahn and Robert D. Yee

Space Biology Laboratory and Department of Psychiatry, and Jules Stein Eye Institute, Department of Ophthalmology, UCLA School of Medicine, Los Angeles, Calif.

Many times drugs prescribed by a physician must be taken at fixed time intervals or at a specific time of the day for maximum effectiveness. When a patient returns for continuing treatment and the physician finds that the prescription has been ineffectual in the management of the disease the lack of information on the medication schedule makes the proper assessment of treatment difficult. In order to assist the physician in the more effective management of medical disease, e.g. with outpatient on self-medication, a record of time at which dosage was self-ministered is important. The Medication Chronolog performs this record-keeping for the patient. This in turn is available to the physician as part of his diagnostics. The genesis of the device was to assist physicians of the Jules Stein Eye Institute in the treatment of glaucoma with out-patients.

The requirements for the device were somewhat biased by its original use, i.e.: small size (king size cigarette pack), light weight, rugged enough to be dropped, capable of being opened by persons with deficient sight, capable of recording the hour of self-medication for a period of at least 2 weeks, and inexpensive. These criteria were met. The device is 10X9X3 cm., weighs under 250 gm., has been dropped and still operates, is being manipulated by patients with defective vision, records for 512 hours which is over 3 weeks, and has electronic parts which cost about $100. (Fig. 1) Prior to constructing this version

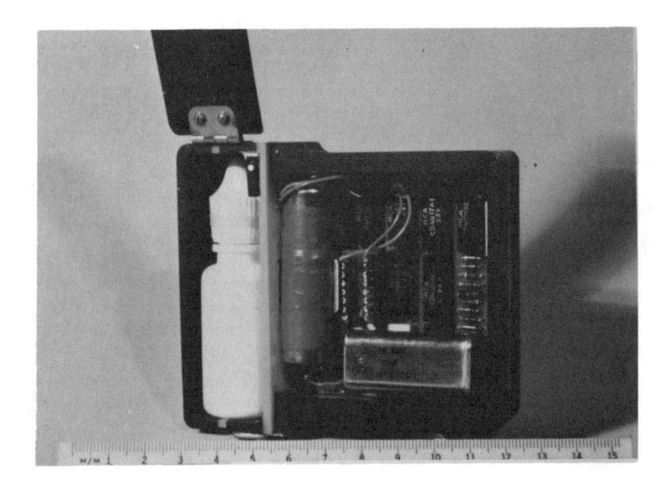

Fig. 1. The chronolog is shown with the lid open and the cover over the electronic circuitry removed.

of the device various recording methods were evaluated. The paper and pencil recording technique works only with well disciplined persons; a clock mechanism driving a paper strip with a stylus for indicating events was too sensitive to mechanical shock; light bulb matrix [24(hr.) x 14(day)] exposing a small piece of film may still be a workable approach; a magnetic strip recorder requires too much power and weight; and various other combinations. All of these methods were rejected either because of mechanical complexity, or high power consumption, or excessive weight. The method adopted uses solid state electronics.

Various solid state designs can implement such data storage. The Complementary Symmetry Metal Oxide-Semi conductor (C-MOS)* family of devices has a power consumption in the microwatt range, has excellent noise immunity, but costs twice as much as Transistor-Transistor Logic (TTL).

The logic consists of a one bit memory which is set to "1" whenever the medication is removed from the device, a clock and divider which interrogates this one bit memory once an hour, loads its status in a 512 x 1 RAM, resets the one bit memory to "0" and increments the RAM address. This is repeated for memory address of 0 to 511. When the memory address is 512 an inhibit signal stops the cycling.

The circuit diagram (Fig.2) is based on RCA C-MOS logic. Two NOR gates (1/2 CD4001) make up the one bit memory which is set by means of a microswitch when the lid is opened. The divide-by-ten counter (CD 4017) does the timing and delays. At the onset of the hour it is

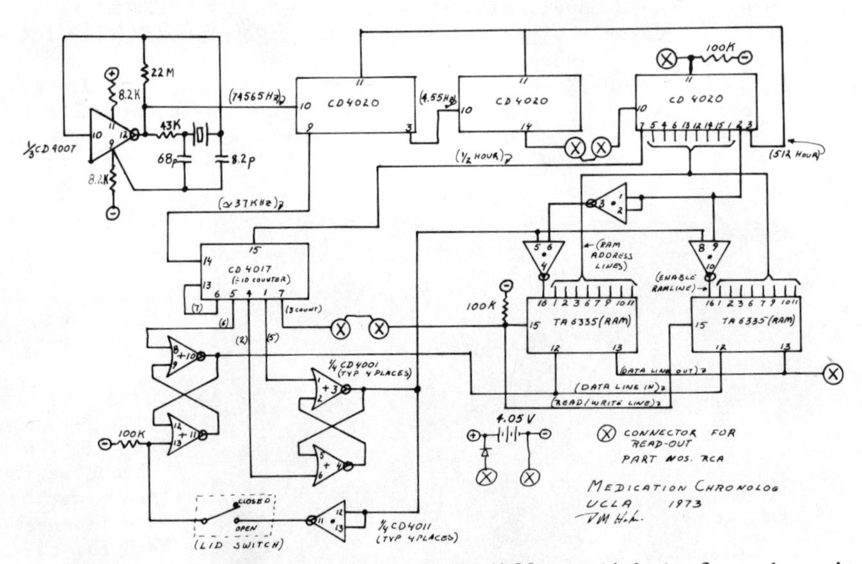

Fig. 2 The schematic shown above differs slightly from the unit presently in use in that an extra IC used for test purpose is not shown.

*RCA Solid State Data Book Series,COS/MOS Digital Integ.Cir.,SSD-203B RCA Solid State, Somerville, N.J., 1974

enabled. At count 2 it enables the random access memories (RAM), at 3 the write signal transfers either a "1" or "0" into the RAM, at 5 the RAM is disabled, at 6 the one bit memory is set to 0 and at count of 7 the divide-by-ten counter is inhibited. This sequence takes less than 20 msec. The RAM (2-TA 5974) were preliminary parts when we obtained them over 2 years ago. Now production models of the 256 x 1 RAM (CD 4061) are available.

By using C-MOS logic we have a worst case power usage of about 250 microwatts. The 4.05V, one hundred milliampere hour cell should last about 2 months. To date the present mercury cell has been in use over 8 months.

In Fig. 2 the symbol ⊗ shows the terminals of the readout and reset connector. A small interface device replaces the dummy connector and allows the device to be "read out" on a strip chart recorder and/or a computer. The interface circuit has its own power supply, variable oscillator and drivers to readout in sequence each of the recorded bits of the RAM. At 5 Hz the data is read out in 2 minutes and both a time line strip and a computer printout (Fig.3) are made, the latter for the patient's chart.

Conclusion The device has been used with four patients to date and has been useful in determining the regularity of self-ministering. Two patients followed the physician's orders, one forgot to take his medication during the weekend and the fourth patient was highly irregular in his habits. These results were important in the assessment of the treatments.

The next generation of this device will be smaller, lighter and the physician will have a small dedicated printer for reading out the chronolog. If the demand is great enough all of the circuitry could be mounted on a single 16 or 24 lead flat pack making the device no larger than the battery needed to operate it.

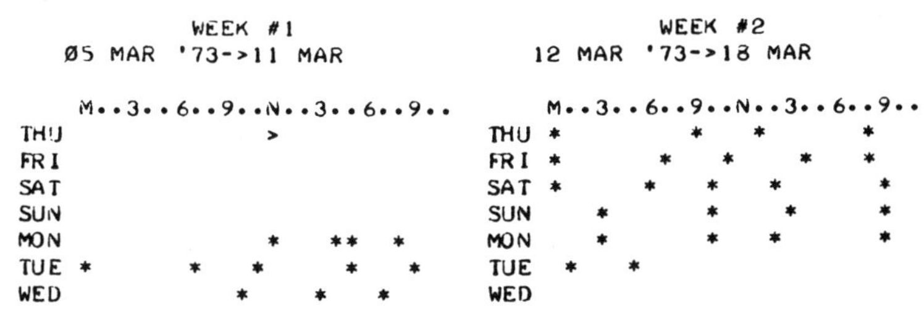

Fig. 3 The patient had an appointment at the clinic on Monday 5 March at noon at which time he received the chronolog which had been started on Thursday 1 March at noon. A week later, Tuesday, 9:00 A.M., he saw his physician and returned the chronolog. His medication was to be taken 4 times daily.

Acknowledgments This research was supported in part by United States Public Health Service Grant # USPHS 5P01 GM 16058.

Biotelemetry II. 2nd Int. Symp., Davos 1974, pp. 74–76 (Karger, Basel 1974)

Digital Memorization of Biological Waveforms

Peider Pinösch, P. Friedli and J.P. Rérat
Institut für Technische Physik, ETH Zürich, Zürich

1. Introduction

Storage telemetry permits the acquisition of biological signals while giving full freedom of movement to the patient or test animal. Waveforms can be stored by an analog tape recorder or in digital integrated circuits with low power consumption (such as the new 1k/8bit C-MOS Random Access Memories (RAM)) [1]. Still, memory capacity remains a severe restriction in the digital case. (This is the price one has to pay for a fully electronic solution which is less cumbersome and, perhaps, more resistant to vibrations and mechanical shocks than an electromechanical analog tape recorder can be.)

2. Digital Memory and Data Compression by Coding

The aim of our project is the construction of miniature digital memories for biological signals. Because of the restriction in memory capacity, suitable preprocessing of the digital signals is of great importance. Starting directly with the digitized signal samples (i.e. with a Pulse Code Modulated (PCM) - form), some sort of data compression [2,3] must be performed. The method chosen depends on the kind of signals one wants to record. In a first series of experiments we considered body temperature and heart frequency of hens and body activity of free ranging deer. In this case a special form of differential coding, the so-called Reduced Delta Code Modulation (RDCM) proved to be advantageous. The next paragraphs will introduce this concept.

2.1 Data compression by advantageous encoding

2.1.1 Pulse Code Modulation (PCM), (Fig.1) Here, each sample is directly the digital conversion of an amplitude. Successive amplitude values have a time difference $t=1/f_s$ where f_s is the constant sampling frequency.

2.1.2 Differential Pulse Code Modulation (DPCM), Fig.1) Instead of encoding each sample itself it is possible to take only the differences between two successive samples. This method often permits reducing the number of bits per word by reducing the dynamic range of the words used. The original signals can be recovered by adding up all difference values. A supplementary bit indicates the sign of the difference.

2.1.3 Delta Modulation (DM), (Fig.1) This means encoding difference values with a single bit only. A logical one means "function increased", a logical zero "function decreased"; "function unaltered" will be translated by a sequence of 0/1/0/1/...

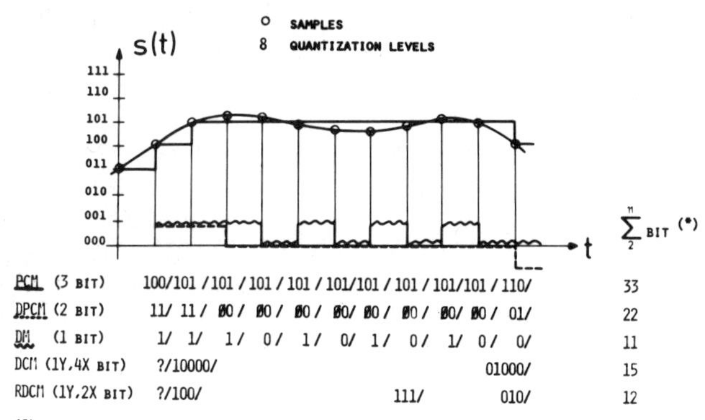

FIG. 1: PCM, DPCM, DM, DCM, RDCM

2.1.4 Delta Code Modulation (DCM) and Run Length Encoding (Fig.1)
In the coding schemes discussed uptil now, the time information was contained in the constant sampling frequency f_s. If several successive samples are equal, the corresponding storage locations possess the same information word. For better utilization of memory capacity, one can encode amplitude (or difference) and time information in a two-part word. The first part Y contains the amplitude (or amplitude difference), the second part X the time until the values of the function have altered. Thus, it is not necessary to store each sample. This encoding scheme is called DCM. For facsimile transmission, there exists an interesting alternative to DCM called Run Length Encoding. Only the X-part of the word is used, indicating thus length of "black" and "white" runs on a (line drawing) picture. At our institute, a method has been introduced which is a combination of DCM and Run Length Encoding: Reduced Delta Code Modulation (RDCM).

2.1.5 Reduced Delta Code Modulation (RDCM), (Fig.1),[2] If long sequences with unaltered amplitudes appear rarely, it is not advisable to adapt the length of the X-word to the longest sequence but rather to use several successive words. A special X-word X_{max} signifies: The amplitude did not change at all in the longest time interval representable by the X-bits. In this case the Y-word is used as a multiplicative constant; when X equals X_{max}, one reads the word as follows: "During Y intervals of X_{max} the function has remained unaltered". Several such XY-words can follow each other, if the run is very long. The following examples illustrate the principle of RDCM. Y: 1 bit, X: 3 bit, X_{max}: 111

```
     Y— X
a) / [0] 010/                function [decreased] after  3 Δt
          3
b) / 0  111/ [1] 110/        function [increased] after 14 Δt
     1 · 7 +      7
c) / 1  111/ 0  111/[0]  010 function [decreased] after 26 Δt
     2 · 7 + 1 · 7 +     5
```

As mentioned before, the most favorable lengths for the X and Y words depend on the time interval statistics of unaltered amplitude values. Optimum word length is calculated in [2]. The following data compression results were obtained with RDCM coding: Body temperature curve of hens and body activity curve of free ranging deer : 30% as compared with DCM, 70% as compared with PCM.

3. Digital Memory Device for Free-ranging Deer

The miniature recorder implemented at our institute stores the body activity of free-ranging deer using the RDCM method. It has been developed in collaboration with the Institute of Wild Animal Research at the University of Zurich. Electrical impulses, proportional to the body activity, are generated by a mercury switch. Four degrees of activity are detected.

Technical data: Sampling frequency approx.: 0,001 Hz (one recording every 15 min. consisting of the average value over the whole interval), dimensions: 30x30x30 mm; memory capacity 1024 bit (storage time 3 days). With higher capacity, IC's just becoming available, a storage time of approximately a month could be obtained. A long range aim is to reach a storage of 1 year (from winter to winter).

4. Other Methods of Data Compression

Other methods of data compression Linear Mapping as especially the Karhunen-Loève-expansion [4,5]) and Redundancy Reduction (e.g. First Order Prediction) [6] are in the process of being investigated. We are applying them to ECG features.

As a first step,"Frame" to "Frame" differences have been formed and encoded (Fig.2). The bit reduction ratio amounts to 50%, as compared with PCM.

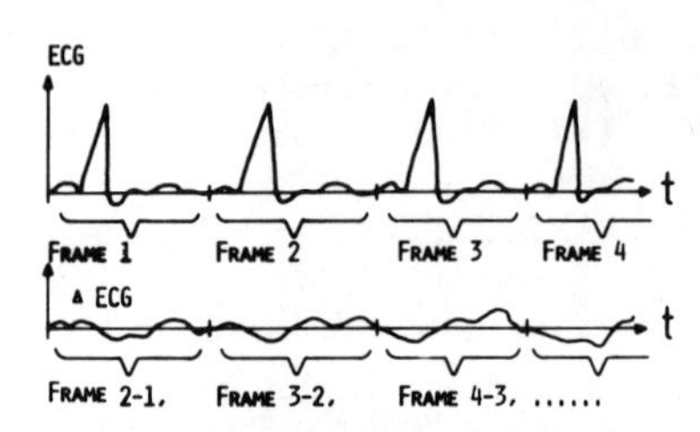

FIG. 2: ECG FRAME TO FRAME DIFFERENCE ENCODING

Acknowledgement

The authors are grateful to Prof.Dr.E. Baumann for his support and to Dr. A. Shah and W. Kraft for their suggestions.

References

[1] HODGES, D.A.: Semiconductor Memories; (IEEE Press, 1972).

[2] FRIEDLI, P.; RERAT J.P.: Digitale Speicherung biologischer Daten bei freilebenden Tieren. AGEN Mitteilungen 15: 35-43 (1973).

[3] PANTHER, P.F.: Modulation, Noise and Spectral Analysis;(McGraw-Hill, 1965).

[4] VAN TREES, H.L.: Detection, Estimation and Modulation Theory; Part 1, (Wiley, 1968).

[5] KROSCHEL, K.: Statistische Nachrichtentheorie; 1. Teil, (Springer)

[6] KORTMAN, C.M. et al.: Redundancy reduction. Proc. IEEE, March(1967).

Session IV

Preprocessing and Reduction of Telemetric Data

Chairmen: *G. Dumermuth and A.A. Borbély*

Biotelemetry II. 2nd Int. Symp., Davos 1974, pp. 78–81 (Karger, Basel 1974)

Processing of Biotelemetry Data

G. Dumermuth

EEG Department, Children's Hospital, University of Zürich, Zürich

With few exceptions, biotelemetry deals with longterm measurements, often from several parallel channels. As a consequence, reduction and automatic or at least semiautomatic processing of the large amounts of data are of great importance. The main goals of automatic processing are (i) quantification, (ii) separation of relevant from irrelevant information and (ii) detection of information usually inaccessible by inspection of ink-written records. Furthermore, automatic evaluation of results is desired.

As most processing methods are performed on digitized data, some words should be said on Analog-Digital Conversion. In a recording-telemetry-processing chain, an important question is how much electronic circuitry can be installed before data transmission. In a freemoving human subject or animal, the amount of circuitry before transmission is limited and digitizing is done after transmission. On the other hand, if the experiment takes place in some moving vehicle, as a satellite, the data are preferably digitized before transmission, (e.g. converted to PCM-format). In the latter case, it will be important to avoid the introduction of redundancy by the digitizing process, i.e. by too much digital resolution or by too frequent sampling, lest the transmitting channel is not overloaded. This is of special importance when transmission has to be done in compressed packages, as from orbiting satellites or from cars or planes moving through silent zones. Table 1 shows the different frequency bandwidths of biological data which, according to the sampling theorem, imply different sampling rates.

After transmission and digitizing (or vice-versa), digital data processing takes place. It is now useful to form three different data categories: (i) ongoing activities with characteristics changing insignificantly or only slowly with time, (ii) transient phenomena, and (iii) activities coupled with some external event (typically, activity evoked by external sensory stimulation). Processing methods generally are different for each of these categories, but there are also common features.

Table 1

Frequency range of biological parameters

Parameter				
Electrocardiogram	0.1	to	150	Hz
Phonocardiogram	5	to	2000	Hz
Ballistocardiogram	0	to	40	Hz
Plethysmogram (volume)	0	to	30	Hz
Pulse waves (peripheral artery)	0.1	to	60	Hz
Blood flow	0	to	20(60)	Hz
Blood pressure (direct)	0	to	60(200)	Hz
Blood pressure (indirect)	30	to	150	Hz
Respiration	0	to	15	Hz
Temperature	0	to	0.5	Hz
Electronystagmogram	0	to	20	Hz
Electrooculogram	0.1	to	100	Hz
Electroretinogram	0	to	20	Hz
Electromyogram (primary)	10	to	2000	Hz
Electrogastrogram (and other smooth muscles)	0	to	0.6	Hz
Skin resistance	0	to	1	Hz
Electroencephalogram	0	to	150	Hz
Intracerebral electric potentials	0	to	5000	Hz

Ongoing activity. As most of the data are functions of time, methods of Time Series Analysis (table 2) are most commonly applied, especially correlation and spectral analysis. It should be kept in mind that correlation and spectral density functions form equivalent counterparts linked by the Fourier transformation. Also most of the various simpler methods listed in table 2 have their corresponding counterparts either in the time or in the frequency domain. Transformation into the frequency domain does, however, not a priori mean that the frequency components have a biological meaning. Fourier transformation is primarily done for analytical convenience. Therefore, other orthogonal systems may also be used for data analysis, as Walsh- or Haar-functions which, however, have the disadvantage of not being time-invariant like the system of sine and cosine functions.

In time series analysis of bioelectric phenomena, the data is considered as being of statistical nature, i.e. as random variables. The results, therefore, are statistical estimates derived from realizations of random processes approximated by the one or the other theoretical mathematical-statistical model.

Table 2

Methods of analysis for ongoing activity

Amplitude histogram:
mean, variance, skewness, kurtosis (excess), test for normality

Correlation and spectral techniques:
auto- and cross-correlation
power spectrum (variance spectrum)
cross-spectrum (amplitude, phase, coherence, gain)
bispectrum (bicoherence, biphase)
complex demodulation

Interval analysis:
period analysis
threshold crossing (zero crossing)
peak or extreme value distribution

Other techniques:
amplitude integration
averaging
analysis of Bernstein
vector iteration techniques

Transients. Analytical methods concentrate on (i) detection of a certain transient wavelet, either partly or completely obscured by technical or "biological" noise, (ii) on analysis of the sequence of occurence of the same transient waveform, and (ii) on measuring in detail the relevant components of the transient under consideration (e.g. EEG complexes). For detection, various methods of pattern recognition are used. However the actual state of the art of automatic pattern recognition is still rather rudimentary, and progress in this difficult field is slow. After detection of the particular transient waveform it may be either analyzed by various direct measuring or analytical procedures, or also considered just as a single unit of a series of events. In the latter case the sequential structure becomes a principal point of interest, and methods of Point Process Analysis are applied.

Activity coupled with external events. Correlation to the sequence of external events (e.g. by stimulus-locked summation) or to a template (template matching), and adaptive filtering are the procedures most frequently used to separate such evoked activity from the non-coupled components.

Further processing. During the last decade it became more and more obvious that most of the analysis procedures mentioned above do not bring substantial data reduction. On the contrary, large amounts of output are produced which have to be further processed. Therefore, the methods above may be considered more or less as preprocessing procedures. The next possible steps are compression into synoptic representations (e.g. contour plots or isometric projections), extraction of relevant parameters (e.g. of spectral peaks) and statistical treatment (e.g. multivariate analysis). This secondary processing brings important problems because of the great diversity of the procedures (partly subjected to personal preferences) and because large computing facilities may be needed.

Biotelemetry II. 2nd Int. Symp., Davos 1974, pp. 82–84 (Karger, Basel 1974)

Storage and Evaluation of 24-Hour EEG-Data Recorded by Telemetry*

A.A. Borbély and W. Brügger
Institute of Pharmacology, University of Zürich, Zürich

The necessity to store and evaluate a large amount of data constitutes a major problem in long-term EEG-experiments. With the present method 24 hour records of electrical brain and muscle potentials, and of other neurobiological parameters, can be stored on analog magnetic tape cassettes, and evaluated with a laboratory computer. The data flow is schematically indicated on Fig.1. The electroencephalogram (EEG), electromyogram (EMG), motor activity, and consumatory behavior of a rat are monitored by telemetry (BORBÉLY et al.,1973). The data (e.g. EEG from 2 different brain sites, neck muscle potentials, and motor activity) are continuously recorded on 4 channels of an ordinary compact magnetic tape cassette (Oxford Instruments, Miniature Analogue Tape Recording System; frequency response 0.2-100 Hz, recording capability 25 hours). Pulses from an external clock can be recorded on one channel as a time-marker. Then the data stored on magnetic tape are played back at 60 times the recording speed into a PDP 11-20 computer where they are converted to a digital form (maximal real-time A/D conversion rate 16 kHz per channel) and stored on disk (RK03, 2.4 million bytes). For visual inspection the digital data stored on disk are read back into the core memory, converted to an analog form, and displayed on an oscilloscope screen. In the current series of experiments, two sleep stages (slow wave sleep and paradoxical sleep) and wakefulness are determined from 24 second blocks of cortical EEG, EMG and motor activity records. The code letters of the diagnosis are entered through the keyboard of a display terminal and stored on digital tape. In addition, the integrated motor activity is automatically computed and stored for each displayed block. In this way 3600 diagnoses of sleep stages or wakefulness, and motor activity values are obtained for a 24 hour period. The data can be plotted out as shown on Fig.2, or subjected to further statistical analysis.

* This study was supported by the Swiss National Science Foundation, grants nr. 3.8790.72 and 3.212.73.

RECORDING

MULTI-
PARAMETER
TELEMETRY

UNRESTRAINED ANIMAL

24 - HOUR RECORDING
ON ANALOG MAGNETIC
TAPE CASSETTES

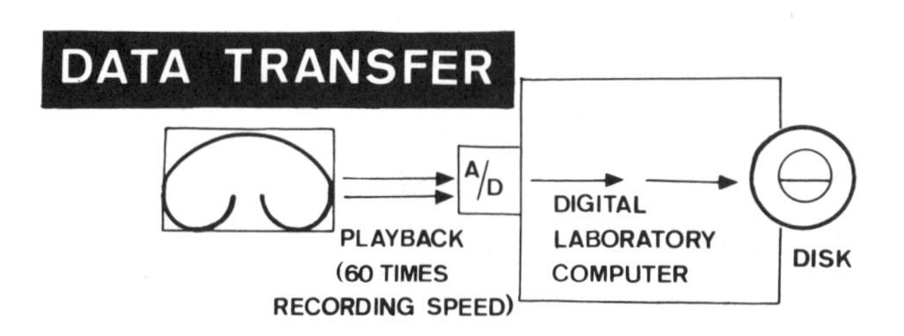

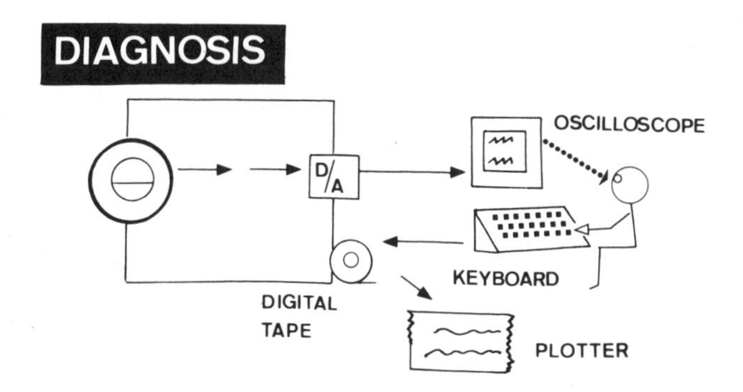

Fig.1 Schematic data-flow for recording, transfer, and diagnosis of neurobiological data.

On the basis of movement criteria, a computer program can recognize unambiguous periods of wakefulness, and thus considerably reduce the time and effort required for the diagnosis. Further programs are being developed to extend the scope of automatic data analysis. In conclusion, analog magnetic tape cassettes consitute a widely available, compact, inexpensive medium for recording data in long-term neurobiological experiments. High-speed playback and A/D conversion make the data readily accessible to computer analysis.

Reference:

BORBÉLY,A.A.,DÄNIKER,M., MOSER,R. and WASER,P.G.: Multiparameter telemetry in neuropharmacological research. AGEN-Mitteilungen 15: 29-34 (1973).

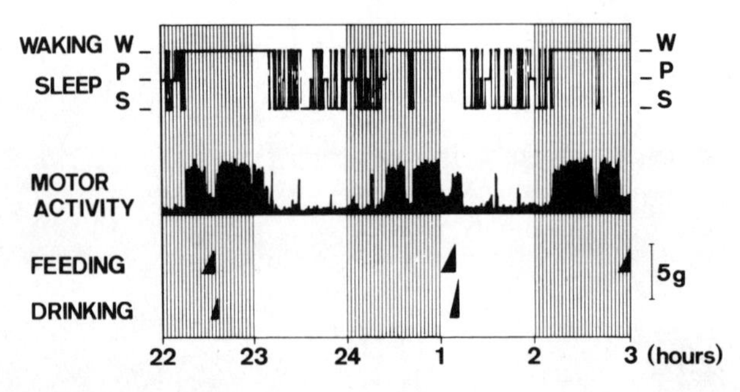

Fig.2 Five-hour section of a record showing sleep-wakefulness, motor activity, and consumatory behavior of a rat subjected to one-hour light-dark cycles (dark periods indicated by shading). P: paradoxical sleep; S: slow-wave sleep. Motor activity plotted in arbitrary units.

Biotelemetry II. 2nd Int. Symp., Davos 1974, pp. 85–87 (Karger, Basel 1974)

Spectral Analysis of Long-Term Electroencephalographic Records Using Band-Pass Filters

Jean-Michel Gaillard
Psychiatric University Clinic of Bel-Air, Geneva

The development of reliable and miniaturized telemetric devices has provided a convenient means for recording bioelectric data of very long duration. The processing of the records however constitutes a bottleneck. Significant information ought to be extracted economically and when possible faster than the recording speed, in order to optimize the utilisation of the telemetric apparatus. This report is concerned with the electroencephalogram (EEG) as it is recorded for instance to monitor the states of vigilance in humans and animals.

The quantification of EEG data may be accomplished in several ways. One of the most suitable for comparative purposes is spectral analysis, which describes amplitude as a function of frequency (WALTER (1973) ; JOY (1971). With the Fourier transform, all the information contained in a short sample of EEG can be extracted. This method requires the signal to be stationary, that is it should not change with respect to time; moreover, due to the high sampling rate necessary for the digitalization, a relatively great computer capacity and a long calculation time are needed. It is nevertheless often useful to get less information, with a lower definition, but in very long EEG samples. The method presented here allows the description of samples of arbitrarily chosen length in a single array of numbers. Since it is a kind of averaging, stationarity may be desired but is not a compelling necessity.

In our studies we use a set of band-pass analog filters of the Butterworth type (SEN Electronique) which is part of an automatic sleep scoring system described elsewhere (GAILLARD and TISSOT (1973). These filters, in Camac (*) modules, are linked via an interface to a small computer NOVA (Data General). The data, electroencephalograms of human subjects or of animals, are processed off-line 16 times faster than the recording time or real time. For simplicity of presentation, all durations or frequencies will be indicated in real time.

The frequencies of these filters are logarithmically distributed between 1,4 Hz and 32 Hz, the band of the EEG where interesting events occurs (8 filters). Four additional filters are used for the detection of artifacts : slow waves below the delta band of the EEG, fast waves of muscular activity sometimes superimposed on the EEG (80 Hz) and 50 Hz activity (twice). The output of the filters is rectified, slightly

(*) Modular instrumentation system for data processing. European Community of Atom Energy (EURATOM), report EUR 4100 F.

integrated and sensed by a discriminator. The sampling rate is determined by a clock incorporated in the system, working at 6,25 Hz. In other words, the rectified and integrated output of the filters is sensed every 160 ms; every filter gives a response of one if its output exceeds the level of the discriminator and of zero in the reverse case. Thus, a twelve-bit word is produced each cycle of the system, with one bit per filter. The level of all discriminators is determined by four bits under control of the program of the computer.

For the special purpose of spectral analysis, the level of all discriminators is incremented by one each cycle of the system. Thus, this level continuously goes over the full range of sensitivity from maximum (level 0) to minimum (level 15), returning to maximum after reaching the minimum. The number of responses for each sensitivity level and for each filter is then calculated as a percentage of the total of possible responses of that filter. Each number expresses the percent of the time during which the filter actually responded, i.e. encountered in the analog signal a frequency corresponding to its band. The full spectrum is printed as an array in which the columns are the different filters in order of frequency and the rows are the sensitivity levels from maximum to minimum.

With this procedure EEG recordings of arbitrary duration (some minutes to several hours) are all described by an array of the same size. It is easy to compare recordings in different subjects, or in the same subject at different periods of time; moreover, recordings of very unequal durations may be compared in the same way. Calculation of the arithmetical mean between subjects and statistical estimation (with a t-test for instance) are also possible.

A plot of the values of a given filter in function of sensitivity level usually exhibits an S-shape. After having discarded the upper values, which may saturate and the lower when they are null, the logarithmic transform of the remaining values yields a straight line. This line may be described by its slope and its ordinate at origin. These two parameters represent another way in characterizing the responses of the corresponding filter. (Fig.)

The method presented here has the advantage of simplicity and rapidity in calculation. In spite of its relatively low resolution, it has proven to be reliable and useful in various kinds of comparisons of EEG recordings.

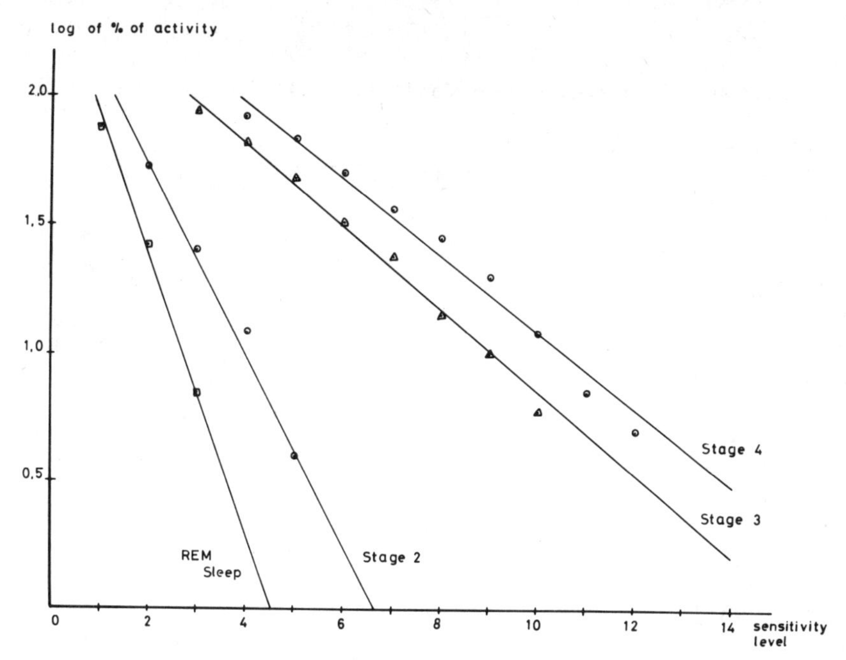

Fig.

Percentage of activity (logarithmic scale) of a filter in the delta band, in function of the sensitivity level of the filter. Comparison between four steady states of vigilance.

References

GAILLARD,J.M., and TISSOT,R. : Principles of automatic analysis of sleep records with a hybrid system. Comput. biomed. Res. 6: 1-13 (1973).

JOY, R.M.; HANCE, A.J., and KILLAM, K.F., jr.: spectral analysis of long EEG samples for comparative purposes. Neuropharmacology, 10, 471 - 481 (1971).

WALTER, D.O. : spectral analysis for electroencephalograms : Mathematical determination of neurophysiological relationships from records of limited duration. Expl. Neurol. 8: 155-181 (1963).

Biotelemetry II. 2nd Int. Symp., Davos 1974, pp. 88–90 (Karger, Basel 1974)

Application of Some Data Compression Methods to ECG Processing

V. Cappellini, F. Lotti and F. Pieralli
Istituto di Ricerca sulle Onde Elettromagnetiche, CNR, Florence

1. Introduction

The digital processing of biomedical signals is of increasing interest for the enlarging amount of data to be processed and for high efficiency that can be obtained.

One important digital processing technique that can be applied to process biomedical signals is represented by data compression. A transformation of this type operates on the signal (in general on the sampled data) reducing the amount of non useful or redundant data: a reduction of the bandwidth, that is required for the transmission of the examined data through the available communication link, can also be obtained [1] .

In this paper we describe the application of some data compression methods to electrocardiogram (ECG) processing * . Algorithms with prediction, interpolation and Fast Fourier Transform (FFT) are, in particular, considered.

2. ECG Processing by Means of Some Data Compression Algorithms

Several data compression algorithms were applied to ECG processing: algorithms with prediction and interpolation, Fast Fourier Transform, digital filtering.

An automatic procedure for data processing with the different compression algorithms was defined by using a standard computer or a minicomputer. After the processing, the following parameters are given: the compression ratio CR (ratio of the original bit number before compression application to the output bit number); rms error, expressed as a percentage of the full scale; peak error, in the same scale as the rms error. Graphical displays regarding the compression performance are given in two separated figures. The first figure shows at the top the original signal as a function of the time; under this diagram a certain number of reconstruction curves is presented, corresponding to a subset of the different parameters actually used for

* Many of these data compression methods were developed for two contracts carried out for ESRO (ESTEC and ESOC) [2].

the analysis (in general the aperture or amplitude tolerance). The second figure is a multiple graph and synthesizes the results of the application of a given algorithm to the same signal with many values of the parameters: the compression ratio CR and the rms and peak errors are, in general, presented as a function of the normalized aperture with respect to the full scale.

2.1. Application of prediction and interpolation algorithms

In the process of prediction, the knowledge of previous samples is used, while in interpolation the knowledge of previous and future samples is utilized. In both types of operation the predicted or interpolated sample is compared with the actual sample: if the resulting difference is within the allowable error tolerance, the actual sample is eliminated. Otherwise the actual sample remains in the compressed data.

We have prediction and interpolation of different orders, depending on the number of samples which are used to define the predicted or interpolated sample.

A zero order predictor (ZOP) defines a predicted sample having the same value as the last output sample with an aperture or amplitude tolerance K specified in different ways. In the ZOP with fixed aperture the dynamic range of the data is divided into a set of fixed tolerance bands with a width of 2K: if the amplitude values of the predicted sample and of the actual sample lie in the same tolerance band, the actual sample is eliminated. An example of ECG processing with this algorithm is shown in Fig. 1: in a) the reconstruction curves are given and in b) the CR, rms and peak errors are reported as outlined above.

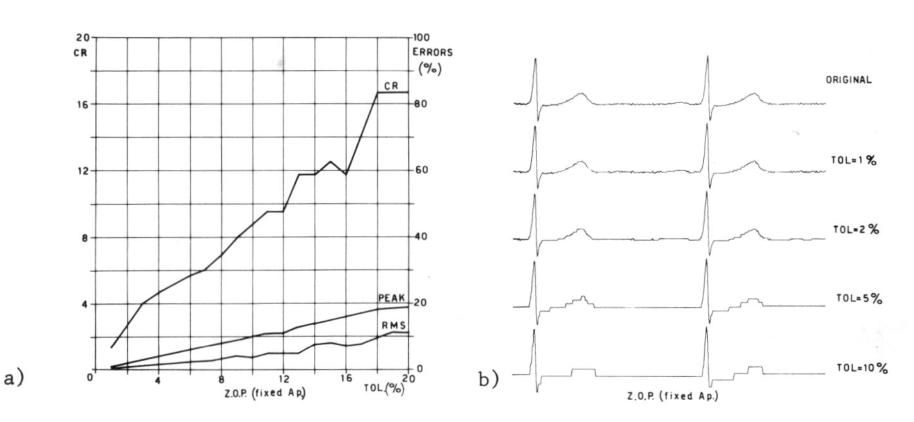

Fig. 1 - Example of ECG (V3) processing with ZOP-fixed aperture: in a) the reconstruction curves are given and in b) the CR, rms and peak errors are reported.

2.2. Application of FFT and digital filtering

FFT was applied to ECG signals to obtain at first a spectral estimation of the signals and after a compressed representation. To this purpose a threshold was generally used eliminating the spectrum amplitude values under a given value (a percentage of the maximum amplitude by not considering the DC- component). Good reconstruction curves were obtained with appreciable compression ratios.

Digital filtering techniques were also widely applied to ECG processing to obtain, as with FFT, a spectral estimation and hence a compressed representation. A special purpose processor was particularly used, performing band-pass analysis and spectral estimation. An analysis up to 16 bands is easily obtained; the spectral estimation is performed by evaluating the rms values of each band output in a given time interval (on 16, 32, 64, 128 or 256 samples). These rms values represent a rough compressed representation of the ECG, that can be transmitted on different data channels with very low bandwidth requirements [3] .

3. Conclusions

Compression ratios in the range of 2 to 10 were easily obtained with the above described algorithms, peak and rms errors remaining in the range of 1 to 7 %.

We have ready the software package to apply other data compression algorithms as differential pulse code modulation (DPCM), delta modulation (DM), Hadamard Transform (HT) and source coding. Original improvements were introduced in these algorithms, generally with suitable adaptive procedures.

From the already obtained results, however, we have the interesting indication that data compression algorithms (also of very simple structure) can be applied to ECG and other biomedical signals to reduce greatly the amount of data and the bandwidth with low errors. In this way, efficient data transmission from different clinical analysis places to a central processing center and a more compact archival of the processed data (efficient data retrieval) can be obtained.

References

[1] KORTMAN, C.M.: Redundancy reduction, a practical method of data compression. Proceed. of the IEEE 55: 253-263 (1967).

[2] Final report for the ESTEC/Contract 902/70 AA concerning A Study on data compression techniques; (Istituto di Onde Elettromagnetiche, Florence, 1971).

[3] CAPPELLINI, V.; EMILIANI, P.L.: A special-purpose on-line processor for band-pass analysis. IEEE Trans. Au-El. AU-18: 188-194 (1970).

Biotelemetry II. 2nd Int. Symp., Davos 1974, p. 91 (Karger, Basel 1974)

An Effective Redundancy Reduction Algorithm for ECG Data

M.J. Kongas
Department of Control and Systems Engineering, University of Oulu, Oulu

The paper describes a redundancy reduction algorithm which is suitable for one-chip microprocessors with limited arithmetic capacity. The input is an ECG signal measured at 3 leads, sampled at 500 Hz, and quantized at 10-bit accuracy. The algorithm is a combination of multirate sampling, digital filtering and variable length coding. A simple nonstationary signal model on which the algorithm is based is also discussed.
The bit compression ratio achieved by the algorithm with negligible signal distortion is about 6. This data compression system is designed for reducing the needed information transfer capacity of a computerized ECG data analysing system.
Editor's note: Final paper has not been received before publication deadline.

Session V

Telemetry of Biomechanical Parameters

Chairmen: *B.M. Nigg and M. Hebbelinck*

Biotelemetry II. 2nd Int. Symp., Davos 1974, pp. 94–96 (Karger, Basel 1974)

Synchronised Photo-Optical and EMG Gait Analysis with Radiotelemetry

Jürg U. Baumann and Rudolf Baumgartner

Neuro-Orthopaedic Unit, Division of Orthopaedic Surgery,
Department of Surgery, University of Basel, Children's Hospital, Basel

Statistically, disorders of the musculo-skeletal system represent the most important cause of long term working incapacity, (WHITE 1973). Correct diagnosis including the type and degree of functional impairment is the foundation for good treatment. Radiographic examination and chemical analysis can provide useful information for such a diagnosis. However, neuro-musculo-skeletal impairment is usually judged mainly on the basis of the physician's visual impressions of movements, particularly gait. More often than not, the patient is examined in a small medical office where only a few consecutive steps are possible and the viewing angle of the examiner is inappropriate.

Gait is a highly automated and complex movement with typical characteristics for each individual. Many attempts for accurate recording of gait disorders have been made in the past. Only recently have electronics made it possible to develop relatively simple procedures providing exact and useful values for the clinician. The latter must be able to correlate the measurements with his visual impressions. In an attempt to obtain better information on gait disorders a method for synchronising electrical data with photographic recordings was therefore developed by the authors.

We are particularly interested in gait in children from their first steps onward. Any attachment to cables had therefore to be avoided. In addition, the equipment carried by the child had to be light and simple. A commercial radiotelemetry system (Narco Electronics, Houston) was adapted to transmit electromyographic potentials and foot -contact impulses. It has the advantage that it can also be extended for the transmission of electro-goniometric and other values. Each emitter including battery weighs 18 g. A separate emitter was used for each EMG-channel and was attached to the body near the site of implantation of wire electrodes according to the technique of BASMAJIAN and STECKO (1962). Up to six channels were employed. The signals were stored on an FM tape recorder. However, kinesiological EMG signals are of little value unless correlated with recordings of the effect of the corresponding muscle action. Considerable information can be obtained from optical recordings during walking. Therefore serial photographs were taken by a Locam[R] camera at a rate of 50/second on 16 mm Kodak Plus-X Negative movie film. The shutter openings were monitored by a photo-diode assembly. These shutter opening impulses were counted and used to provide marking impulses for each picture on the tape. Every tenth impulse was identified by means of a binary code. Thus 999 consecutive camera opening impulses could be identified on the tape. This method of recording permits demonstration of gait or any other movements either in cinema projection or as a gait-phase-photogram. Cine-projection provides the impression of movement which can not be replaced completely by any other information on normal and abnormal motor performance. The gait-phase-photogram (Fig. 1) is assembled from enlarged prints of 16 mm film negatives and write-outs from the correlated electrical recordings on magnetic tape by a UV writer at 2 m/s and represents one gait cycle. The individual pictures were taken with an intervall of 40 ms and an exposure time of 6.6 ms.

In order to obtain pictures from several steps, the camera was mounted on a travelling trolley and pushed on rails parallel to the walking patient. Views from the side, back and front were taken consecutively.

Reviewing the results of 72 examinations in 60 patients and compairing them with the force-plate measurements of BRESLER and BERRY (1951) as well as the pioneer-work of LEVENS, INMAN and BLOSSER (1948) on transverse rotation between lower limb segments and the trunk, the following conclusions could be drawn:

The conventional terminology of gait phases should be revised. Up to now the terminology has been based exclusively on the relationship between the foot and the floor. Each step was divided into stance- and swing-phase according to the presence or absence of floor contact by the foot. However, the results of these examinations show that the transition from stance to swing starts long before the foot actually leaves the ground. At 50% of the gait cycle the hip joint changes from extension and inward-rotation to flexion and outward-rotation and thus to the phase of supported swinging. The following modification of gait cycle terminology is proposed:

1. Transition swing - stance: 0-3% of gait cycle.
2. Braking and transverse rotation: 3-13% of gait cycle.
3. Support predominant: 13-27% of gait cycle.
4. Push predominant: 27-50% of gait cycle.
5. Supported swing: 50-65% of gait cycle.
6. Acceleration of free swing: 65-85% of gait cycle.
7. Deceleration of swing: 85-100% of gait cycle.

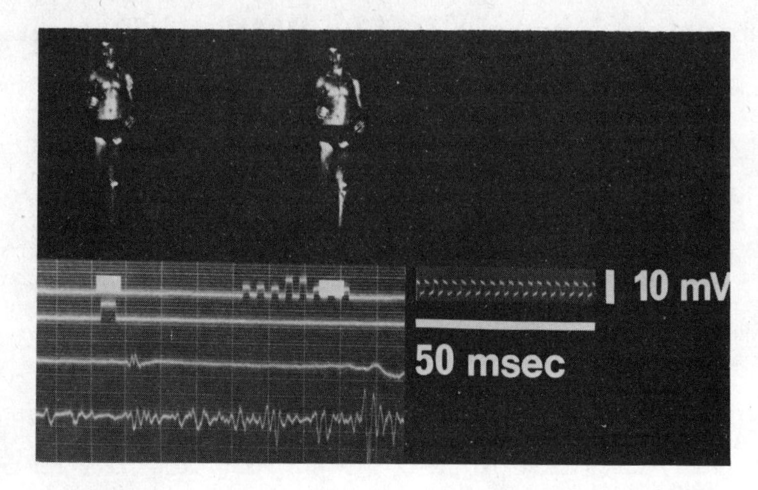

Fig. 1 Gait-phase-photogram

1 - Film exposure marks and counting
2 - Heel contact
3 - EMG semimembranosus
4 - EMG semitendinosus

Literature

BASMAJIAN, J.V. and STECKO, G. A new bipolar electrode for electromyography. J.Appl.Physiol. 17: p. 849 (1962).

BRESLER, B. and BERRY, F.R. Energy and power in the leg during normal level walking. Prosth.Dev.Res.Proj. Series 11, 15 (Univ. of Cal., Berkeley 1951).

LEVENS, A.S., INMAN, V.T. and BLOSSER, J.A. Transverse rotation of the segments of the lower extremity in locomotion. J.Bone Jt. Surg. 30-A: p.859 (1948).

WHITE, K.L. Life and death and medicine. Scient.Amer. 229, Nr.3: p. 23 (1973).

This work was supported by a grant from the Swiss National Research Fund.

Biotelemetry II. 2nd Int. Symp., Davos 1974, pp. 97–99 (Karger, Basel 1974)

A Radio Controlled Light-Pulse Equipment for the Photographic Measurements of Human Body Movement

Raymond Grandjean and Peter A. Neukomm

Laboratorium für Biomechanik, ETH Zürich, Zürich

Measurements with light-pulse equipment is a method similar to the already familar film-analysing, but the photographs we get [instead of films] contain the desired information in a form that is condensed and easy to evaluate.
This is the basic idea:
When a light is somehow moved in a dark room, it will appear in the form of a continuous trace on a photograph, taken with a long exposure time. Pulsing of the light-emission will result in a dotted line, and this contains excellent information on the light's velocity if the distance Δs and the time Δt between two dots are known. The velocity is then given by $\Delta s/\Delta t$, matching closely its real value ds/dt for a short Δt. For measurements in Biomechanics, of course, the pulsed light-emitters are to be attached to a human body.

When we projected the new equipment, it was our aim to offer easy and comfortable handling to the user without sacrificing flexibility or information capacity. This task was fulfilled by Mr. Peter A. Neukomm and under his supervision and technical advice, three groups constructed the necessary parts of the entire outfit.
First, the light-emitting assemblies were developed, each containing a miniaturized receiver, an accumulator and a driver stage for the bulb. The second part is the command unit, with which a suitable digital program may be selected and radio-transmitted. The power supply again is an accumulator.
Finally, the charging unit for the other parts was constructed. It is combined with a radio-controlled reference-scale with 11 bulbs from 10 to 10 cm. Charging of the light-emitting assemblies is done simply by unscrewing the bulbs from their sockets and by plugging the latter into a hole of the charging unit.

When measurements take place, the proceedure is as follows:
- the receivers are attached to the test person's body,
- both the reference scale and the test person are placed at exactly the same distance from the camera. The scale shall appear on every photograph in order to allow measurements of Δs,
- a digital program is selected on the command unit,
- the program is transmitted, and the diaphragm of the camera remains open while the test person moves.

LIGHT PULSE EQUIPMENT

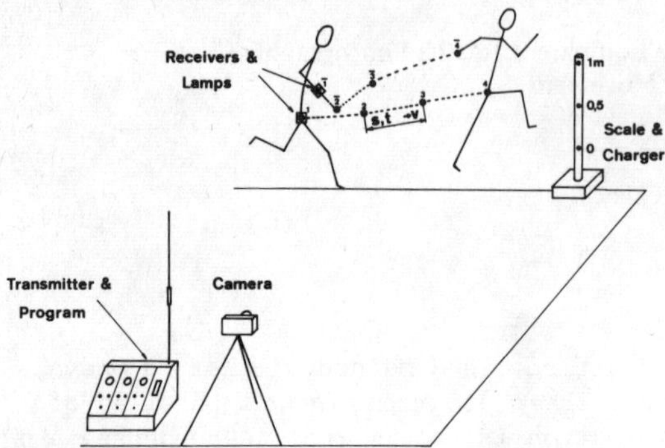

Fig. 1

The principle of the light-pulse equipment

To enhance information quality in the photographs, the following block diagram was realised in the command unit.

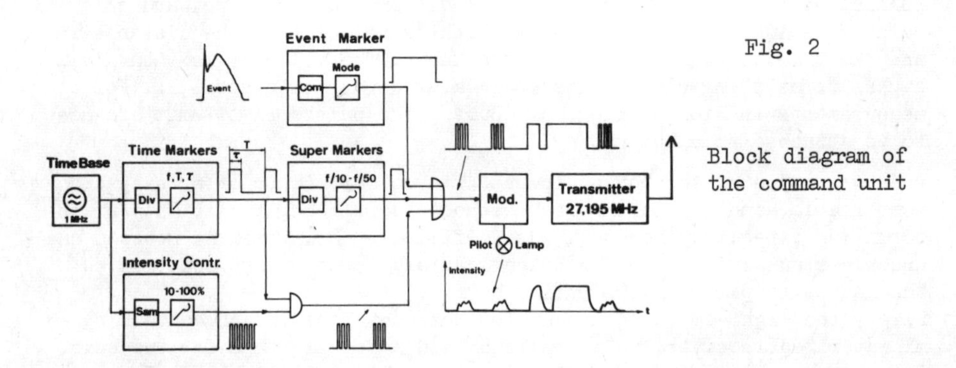

Fig. 2

Block diagram of the command unit

The time base gives a signal of 1 Mcps to the divider stage, where the frequency and pulse length of the time markers are selected. [f = 1 cps to 100 cps in 1, 2, 5 decadic steps]
The super markers control takes this frequency over and reduces it to 1/10, 1/20 or 1/50, also permitting choice of a suitable pulse length. The super markers result in the form of longer and brighter light spots, periodically interrupting the normal time markers. This provides much easier relative position identification of the different body parts.
In the intensity control, the time markers can be sampled. A filament lamp has low pass characteristics for its light emission; the time constant is given by its thermal resistance and storage capability. Therefore, a quick sampling reduces the lamp's brightness, while a lower frequency [in our case maximal 100 cps] is still reproduced as a dotted line.

Special events, e.g. the moment of touching the ground after a jump, can be emphasized by the event marker control that is triggered externally if required.
The three signals are "OR" connected and modulated to the transmitter simply by on-off switching of its output stage.

We wanted all the equipment to be independent of main circuit connections, so a 10 V accumulator feeds the transmitter. Since the logic functions turned out, however, to be quite complex, power consuming problems would have arisen by using a TTL-family. Fortunately, the complementary MOS-family was introduced in Switzerland when we began our work. With its extremely low power consumption and other very good features, it applied perfectly to our equipment.

The light-emitting assemblies are the result of thorough studies on miniaturizing problems. The strip line antenna is a part of the two printed circuit boards, that form the top and the bottom of a sandwich construction with the accumulator and the electronic components inside. The electrical layout is conventional, consisting of a crystal oscillator, an intermediate frequency filter stage and the bulb driver. Only separate elements were used, as integrated stages required the same space for their external components, and power consumption would have been excessive [20 mA instead of 3 mA for the high impedance circuit with separate devices].

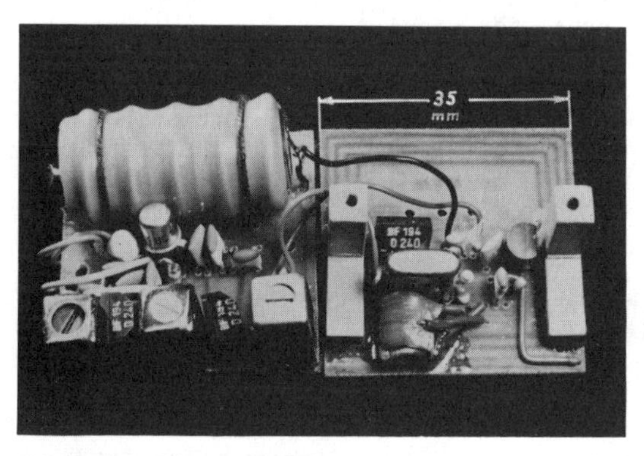

Fig. 3 RC-receiver

There is still one problem to discuss: the equipment was ready for use, but the room where we worked had to be entirely dark as long times of exposure were inevitable. As a consequence, we lighted the room with special, blue fluorescent lamps. Their spectrum is so different from the spectrum of the bulb that it is filtered out easily with a red filter.

Today, first tests have been made and the results are encouraging. The equipment applies very well to all quick and non repetitive movements and these are fortunately extremely frequent in gymnastics and sport.

Acknowledgment
We are grateful to Prof. Baumann, head of the Inst. of Applied Physics, where Mr. K. Walser and Mr. H. Walter developed the miniaturized receiver during a regular study work. Ref. Semesterarbeit 1973, Inst. für Technische Physik, ETH.

Biotelemetry II. 2nd Int. Symp., Davos 1974, pp. 100–102 (Karger, Basel 1974)

An Optoelectronic Instrument for Remote On-Line Movement Monitoring

Lars-E. Lindholm and Kurt E.T. Öberg

Dept. of Applied Electronics, Chalmers University of Technology, Göteborg,
and R & D Department, Een-Holmgren Ortopediska AB, Uppsala

A new optoelectronic instrument capable of accurate remote recordings of movements, e.g. limb movements, has been developed. The instrument consists of a special photodetector which provides two electrical analog outputs, specifying the individual x and y positions of input light-spot signals in a fixed internal coordinate system. Small light-emitting diodes are used to provide the input spot signals.

The number of diodes that can be simultaneously used is well above one hundred. The instrument has a bandwidth of 500 Hz for each channel. The upper limit for the range of the instruments (the distance between the diodes and the detector) is about 20 meters with a 50 mm lens, but can easily be increased by using more powerful diodes.

The resolution of the instrument is about 1 part in 1000 and the maximum deviations from the linearity are about 0.2 per cent.

Compared to high-speed filming the instrument has the advantage of giving an immediate result, without the need for processing and measuring tens of thousands of film frames.

The instrument can be used in a variety of medical and industrial applications. One application of the instrument is to provide a mathematical model of human gait with input data, which determines the instantaneous position of the body during walking. The measurement is carried out by affixing light-emitting diodes to the body. The position of the diodes, and thus the position of the body, is determined by the instrument output. The instrument is connected on-line to a minicomputer, where the processed data can be stored and punched on tape for further calculation later on or immediately, through the mathematical model-program, be processed to the form of information that is wanted.

The model consists of a two-dimensional mechanism with seven jointed segments corresponding to foot, shank and thigh for both legs and the trunk including head and arms.

A Fortran IV program is written to calculate the rest of the necessary coordinates and angles through the relations between the body segment parameters and the input data. It calculates velocities and accelerations through numerical derivation and hence the moment and force function in the joints can be calculated through solution of the Newton's equations for each segment.

A lot of information from input data and calculated data is then available in the computer and can, through written routines, be obtained as diagrams and tables in varying forms.

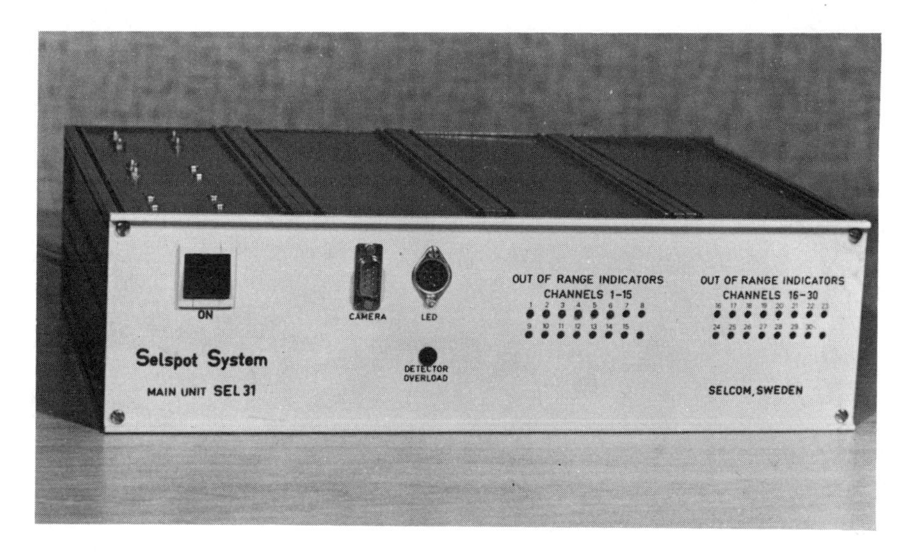

Figure 1: This picture shows the "main unit" for the SELSPOT. This device contains the necessary electronics for the processing and distribution of the signals from the camera.

Figure 2: This picture shows the SELSPOT camera and the "LED Switching Box" which contains the LED switching electronics. This box is carried by the subject being measured. The box can supply up to 30 LED:s.

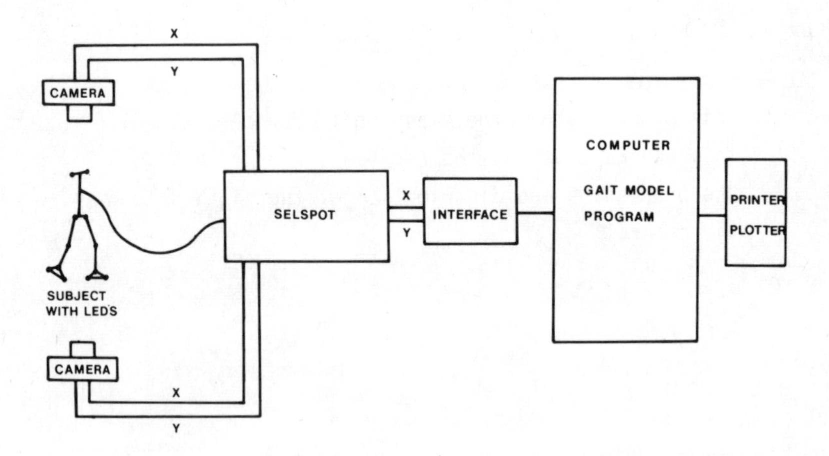

Figure 3: This picture shows a setup for on-line measuring and processing of gait.

Biotelemetry II. 2nd Int. Symp., Davos 1974, pp. 103–105 (Karger, Basel 1974)

Telemetry and Training Control of Swimmers

R. Deroanne, M. Leloup, F. Pirnay and J.M. Petit

Institut E. Malvoz, Province de Liège and Médecine et Hygiène Sociales, Université de Liège, Liège

When a coach wants to increase the resistance of an athlete, he generally determines distances and duration of training according to the performances reached by this athlete in competition. However, it is difficult to find beforehand the intensity of the effort imposed and to verify its level during training. The means used by the coach are his own appreciation of the fatigue state of the athlete and also the athlete's own impressions.

The pulse rate measurement is a more objective proof that can be used very easily by means of a telemetry system.

In this study, the results of heart rate telemetric recording during fractionned training of swimmers are compared to their ergospirometric data obtained in laboratory. The subjects are top swimmers (6 boys and 4 girls). In the laboratory, the Maximum Steady State and the Maximum Oxygen Consumption (PETIT et coll, 1962) are measured by means of an open circuit method.

The results of Maximum O_2 consumption obtained on the boys (mean 60,1 ml O_2/min/kg) as well as on the girls (mean 43,4 ml O_2/min/kg) give proof of a very good physical fitness. The results of the heart rate telemetric recordings during training periods demonstrate very high intensity of work (table 1). Furthermore, the evolution of the heart rate of each swimmer during his training session shows a perfect correlation between the tolerance of the work intensity and the heart rate corresponding to the Maximum Steady State. Fig. 1 shows the heart rate evolution of a swimmer who tolerates the training intensity and fig. 2 shows the evolution of the same parameter by a swimmer who does not tolerate the training intensity.

Heart Rate Individual values measured at the Maximum Steady State and the mean value measured at the end of the courses.

Subject	JG	MD	GP	DC	JL	ML	DD	FC	TB	MV
MSS	197	206	190	179	189	165	195	190	187	192
Mean at end of courses	189	202	183	168	189	190	197	177	195	192

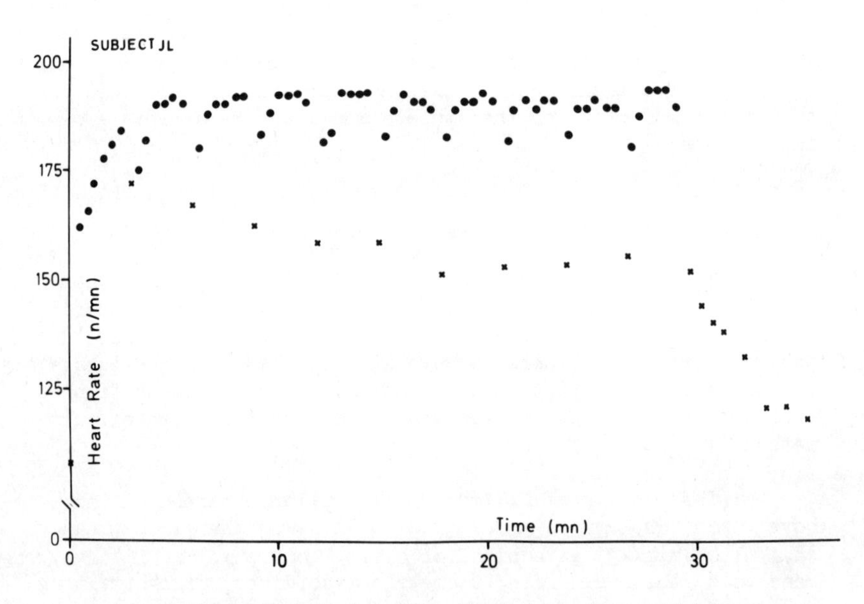

In this study, the absolute values of heart rate are often much higher, (mean 200 beats/min) than that described by MAGEL and FAULKNER (1967) or MAGEL (1971). At the end of recovery period, the heart rate reaches high values even for swimmers who easily tolerate the training intensity. The limit of 120 beats/min generally fixed before a new effort starts (JUGE and SPRECHER, 1960) is always surpassed.

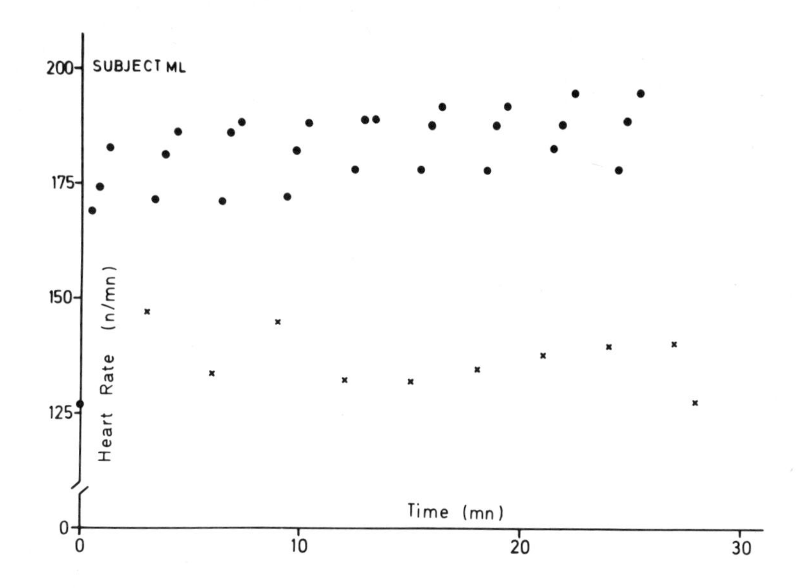

In some other cases, it has been possible to put in evidence a lack of motivation and collaboration from the swimmers.

In conclusion, the heart rate telemetric measurements inform us on the absolute and relative intensity of the effort imposed, on the training type and on the subject's motivation.

JUGE A. et SPRECHER P. Entraînement fractionné. Education Physique et Sport, 1960, 50, 26-28.

MAGEL J.R. Comparison of the physiologic response to varying intensities of submaximal work in tethered swimming and treadmill running. J. Sports Med. 1971, II, 203-212.

MAGEL J.R. and FAULKNER J.A. Maximum O_2 uptake of college swimmers. J. Appl. Physiol. 1967, 22, 929-933.

PETIT J.M., DELHEZ L., DAMOISEAU J., BELGE G., COLLEE G. et DEROANNE R. Mesure simplifiée de la consommation maximum d'O_2, test du degré d'entraînement. Comm. au Congrès Int. Ed. Phys. Liège mai 1962.

PETIT J.M., PIRNAY F., DEROANNE R. et BOTTIN R. Mesure simplifiée du régime stable ventilatoire maximum au cours de l'exercice musculaire. Acta Tub. Belg. 1966, 57, 71-82.

Biotelemetry II. 2nd Int. Symp., Davos 1974, pp. 106–108 (Karger, Basel 1974)

Telemetry of Dynamic Forces in Endurance Sports

J.A. Vos, H.P. Kimmich, J. Mäkinen, H.J. Ijsenbrandt and J. Vrijens
Dept. of Physiology, University of Nijmegen, Nijmegen

At the International Symposium on Biotelemetry (1971) examples of measuring maximal static strength by means of strain gauges and one possibility of measuring dynamic force of both legs while rowing were given.Men's interest in strength development has completely changed in our society.Most of us think we are no longer dependent upon physical strength and endurance in our daily circumstances,so that you will find the need forof strength development by training programs and measuring muscular strength only in sport and rehabilitation.

Method: We developed a battery of selected,well standardized tests in which we measured maximal static strength of different large muscle groups,in order to receive more information about the " strength pattern " of the human body.All measurements are recorded by means of strain gauges.A maximal force is stated as the peak value of three well-exerted trials. Description of the battery was published in the proceedings of the Biotelemetry Symposium.During the last two years we developed standards for various sportsgroups and untrained people.

Analysis of the rowing or canoe-movement (kayak) from the shore has to be called difficult because the performance has to be done on water and this situation is more or less invisible from the shore.We measured dynamic strength while rowing of eight well trained rowing girls and four well trained men in kayak.The telemetry system we used is described by IJSENBRANDT (1972).We used three channels for measuring force,namely one channel for 'toes force ';one channel for 'heel force' and one channel 'torque force'.The strain gauges glued on the aluminum footplate are connected as a Wheatstone bridge and the electric current due to strain was amplified and transmitted to the recorded set-up on the shore.The same method was followed in measuring the strength on the paddle while canoeing (kayak),only in this case we glued the strain gauges on wood and used two instead of three channels of the telemetry system.

Results: The curve in Figure 1 shows an exemple of the force while rowing of a well trained rowing girl.Measurements are done at sprint stroke pace of 34 strokes per minute.With all eight girls we measured moderate training stroke pace, (25 strokes per minute);competition stroke pace(31 str/min.) and sprint stroke pace (34 str/min.).The results are shown in Table I.

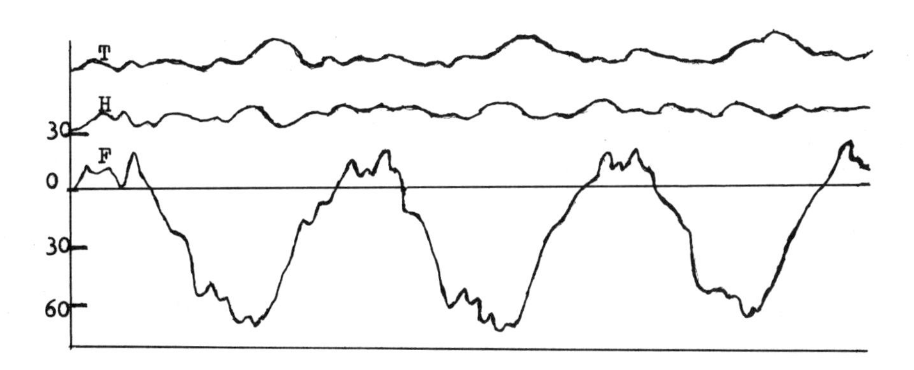

Figure 1. T = 'torque force'; H = 'heel force'; F = 'toes force'. Results are given in kgf.The measurements are done with M.T.(rowing-girl) at sprint stroke pace of 34 str/min.

Table I

FORCE ON TOES:(kgf).*

	A.	B.	C.
1.N.Ca.	71 - 74 - 86	84 - 80 - 80	90 - 48+) - 80
2.C.Du.	58 - 55 - 52	74 - 71 - 71	71 - 61 - 58
3.N.Du.	61 - 55 - 58	80 - 80 - 74	80 - 77 - 77
4.M.To.	51 - 61 - 61	64 - 68 - 64	48+) - 71 - 74

A = average training stroke.(25 str/min.).

B = competition stroke.(31 str/min.).

C = sprint stroke.(34 str/min.). +) = 'failstroke'

The results of measuring the average force at the paddle while canoeing (kayak) are demonstrated in Table II.

Table II.

FORCE ON PADDLE:(kgf).*

	A.	B.	C.	D.
1.P.S.	15 - 16	37 - 40	35 - 40	40 - 42
2.D.B.	-- --	31 - 36	31 - 34	33 - 37
3.J.B.	-- --	31 - 34	32 - 30	35 - 32
4.P.H.	16 - 17	33 - 37	32 - 36	33 - 37

A=average training pace. B= start pace.(left and right)

C=Pace 500 meter competition. D= Sprint pace.

Discussion:What happens above the water is very intensively watched by coaches and described in many books and articles. What happens under water was our interest;information about the strength while rowing or canoeing will be necessary to arrange a well balanced conditioning and weight training program.

Literature:

IJSENBRANDT,H.J.B., KIMMICH,H.P. and VAN DEN AKKER,A.J., Single to seven channel lightweight telemetry system. Biotelemetry Proc.Int.Symp., Meander, Leiden, The Netherlands.(1972)

*kgf = kilogram force or kp

Biotelemetry II. 2nd Int. Symp., Davos 1974, pp. 109–111 (Karger, Basel 1974)

Physiological Aspects of Synthetic Field Tracks

Foot-Motions During Supportphase of Various Running Types on Different Track Materials

R. Haberl and L. Prokop
Österreichisches Institut für Sportmedizin, Wien

1. Introduction: Synthetic athletic tracks are a reality in today's sport practice and have to be coped with in some way by athletes, trainers and physicians. These tracks are said to improve performances in various light athletic disciplines spectacularly. Similar arguments have, however, been voiced by synthetic track critics, referring to excessive "hardness", tissue-incompatible elasticity and the injuries and chronic damage directly provoked by the above characteristics.

A representative inquiry among European top athletes resulted in the first objective pointers (1). Therefore the socalled "Tartansyndrome" is a reality and this fact is well known to the concerned athletes, trainers and physicians; it results in acute, chronic or periodic irritant states, particularly in bradytrophic tissues. These in their turn give rise to painful alterations in the ankle-joint, in the musculature of the lower thigh and the Achilles tendon, as well as in the periost of the tibia and the knee cap and later in the vertebral column that may cause athletics incapability.

2. Methods: The symptomatology clearly established in the course of the experiments points etiologically to a specific effect of the synthetic track material on the motional apparatus during the supportphase, when there is frictional contact between the running shoe and the ground. It was therefore necessary to try and obtain evidence of specially measurable differences by means of a comparison between standardized forms of motion on various surfaces.

A mechanics model was used for the interaction between the ground, feet and body and mathematically treated. The vibration differential equation was soluble under special initial conditions.

Because very little complete data of relevance to the elasticity and absorption in human tendon and muscular tissue could be found in literature, a great many preliminary experiments had to be undertaken. Since it was initially necessary to undertake measurements without influencing the motional processes to be measured, high speed cinematography was found to be the method of choice (2). It showed a basic trend towards damped mechanic vibrations. The initial amplitude of these vibrations was found to be the most favorable characteristic for the evaluation of different surfaces. To measure that, a Biotelemetry system was specially adapted and a sprinter of high technical standard served as a test person (3). His track-shoes were fitted alternatively with a piezoelectric acceleration-transducer in the rear row of pegs either left or right. This device added approx. 50gr. to the weight of the shoe. The signal so recorded was adapted to the input of a single channel telemetric transmitter via a miniature impedance transformer in the athlete's belt and transmitted radiotelemetrically. On stationary reception and demodulation, the signal was UV-recorded. It was additionally possible to label the beginning of the supporting phase externally. For technical reasons it was decided to test only a single representative from each of the important surfacing groups:

a) Cinder track and Turf track.
b) Porplastic, representing all bitumen-gravel and bitumen-rubber tracks.
c) 15mm Regupol, representing all elastic tracks on a rubber-, PUR-, PVC- or similiar basis.

3. Results: Two measuring stages were found to be necessary due to the proportions of the measuring signal, the signal being registered 1:1 or respectively approx. 1:9 reduced. (Fig. 1) This was necessary since the acceleration representing the dampened vibrations was smaller than 1g, while the shock impulse peaks at the beginning of the supporting phase were found to be within the 30g range. Both measuring stages together resulted in a complete picture of acceleration behaviour and, after double integration, also of the foot's movement in a vertical direction during the supporting phase.

3.1. Shock vibration: Based on a correction of the mass ratio of the shoe, changed due to the pick-ups, we obtained a frequency between 80 and 110 cycles according to running speed and elasticity

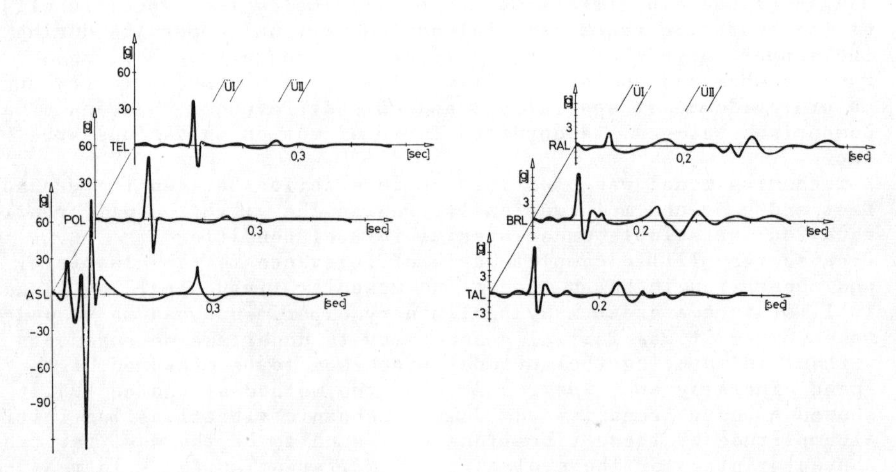

Fig. 1: Acceleration diagrams of a foot on Cinder track (TEL), Bitumen-rubber track (POL), Bitumen-gravel track (ASL), Turf track (RAL), Rubber track (BRL) and PUR track (TAL).

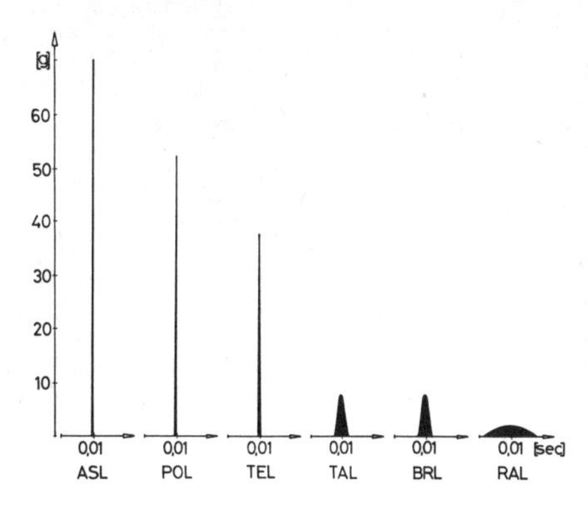

Fig.2:Initial decelerations. Meaning of denotations see Fig.1

of the track material,thus confirming the mechanical model.We also obtained pointers as to the dependency of the frequency on the supporting force during the supporting phase and possibly also on the damping of the track material,a circumstance to be taken into account in the mathematical model used.No harmonic components could be shown with some "very soft" track surfaces,e.g.turf areas,and some "very hard" surfaces,such as bitumen-gravel and bitumen-rubber tracks.

3.2.Initial deceleration:In toto the initial negative acceleration was found to lie between 2g and 80g,there being clear differentiation into three groups:The group with values above 25g,including especially all bitumenbound tracks and Cinder track.The group with acceleration values between 5g and 25g and a focal point between 8g and 12g,including in particular the elastic tracks on various plastic bases and finally the group with values below 5g, for instance the turf areas.

4.Discussion:The results obtained impressively confirmed the hypothesis concerning the origins of the "Tartansyndrome".With elastic track materials it was possible to prove a doubtlessly effective shock vibration,which was almost completely lacking on conventional Cinder tracks.On bitumenbound tracks,where there was no elastic vibration,the initial shock acceleration was found to be greatly increased in comparison to conventional surfacing and this was the origin of the "Tartansyndrome" there.The size of initial shock acceleration was found to be in full agreement with the athlete's subjective judgement as to whether a certain track was to be characterised as "too hard" or "too soft":
"Too hard" tracks show accelerations of more than 30g,"good tracks" approx. 10g to 30g,and "too soft" tracks below 10g.

5.References:

HABERL,R.;PROKOP,L.:Die Auswirkungen von Kunststoffbahnen auf den Bewegungsapparat
(1) 1.Mitteilung,Ö.J.f.Sportmed.2/72 p.3-19
(2) 2.Mitteilung,Ö.J.f.Sportmed.3/72 p.3-32
(3) 3.Mitteilung,Ö.J.f.Sportmed.4/72 p.3-12

Biotelemetry II. 2nd Int. Symp., Davos 1974, pp. 112–114 (Karger, Basel 1974)

Continuous Acceleration Measurements in Humans During Locomotions on Different Surfaces

E.M. Unold, P.A. Neukomm and B.M. Nigg
Laboratorium für Biomechanik, ETH Zürich, Zürich

The laboratory of Biomechanics in Zürich investigates accelerations of the human body caused by different movements on different surfaces. Therefore measurements were made in walking and running, in artistic gymnastics and in skiing. It is interesting to observe from this what forces influence the human body. But these forces can not be measured directly with our facilities, so we measured the accelerations.
Fig.1 shows a schematic diagram of the Telemetry-System used for all our measurements. In the top of the picture is the common part of a 7 channel Telemetry-System. The lower part is the radio-controled-Remote-System. It consists of

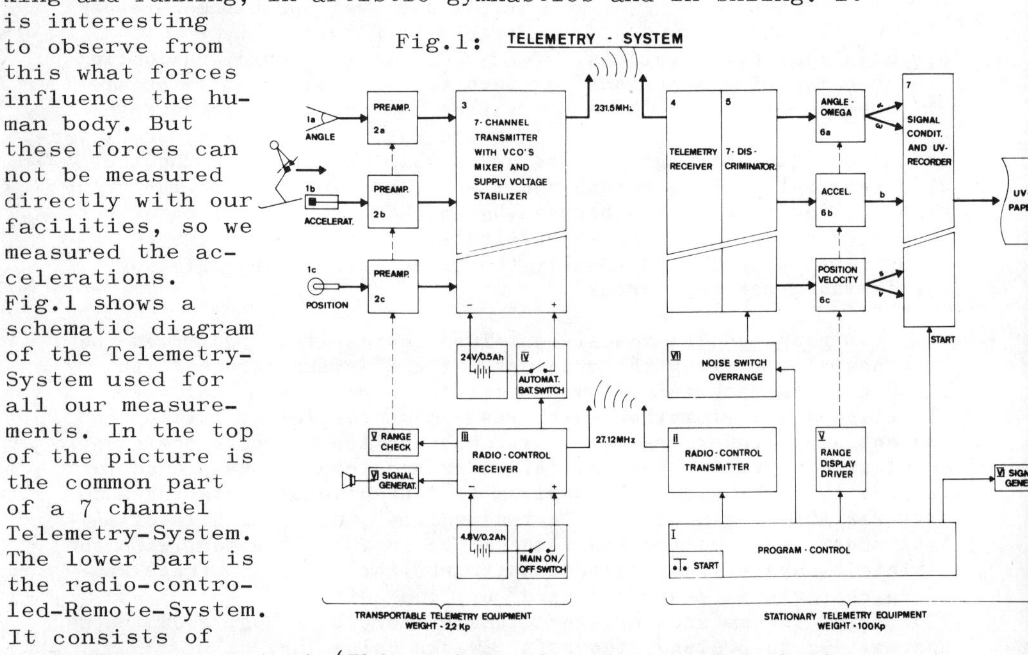

Fig.1: TELEMETRY · SYSTEM

- the Program-control (This drives the starter and checks the data transmissions.)
- the radio-control-transmitter
- the radio-control-receiver
- the automatic battery-switch (The signal transmission runs only during the moment of data-recording.)
- the signal generators (They produce the start signal for both testsubject and operator.

For all the measurements three accelerometers are attached to the subject, one at the shin-bone, one at the hip and one on the head at points where there is not much muscle and fat. All three accelerometers measure in the longitudinal direction of the body. The weight of the telemetry-transmitter is approximately 2.5 kp so that it can carried at the back for many kinds of movements.

Measurements in walking and running

Tab.I shows all tests made with different types of locomotion on different surfaces and with different kinds of footwear. There are only small differences between the single accelerations at the hip and on the head, but there are great differences between the values at the leg. This means: As great as the accelerations at the shinbone are, they shall be dampened in the lower extremities so that the spine is not influenced. This is certainly a factor which causes more and more injuries to the lower extremities. The greatest accelerations at the leg appear in the tests on asphalt and synthetics. Therefore a synthetic floor is not very good for the health. So it would be better to train only a minimum of time on this type of surface.

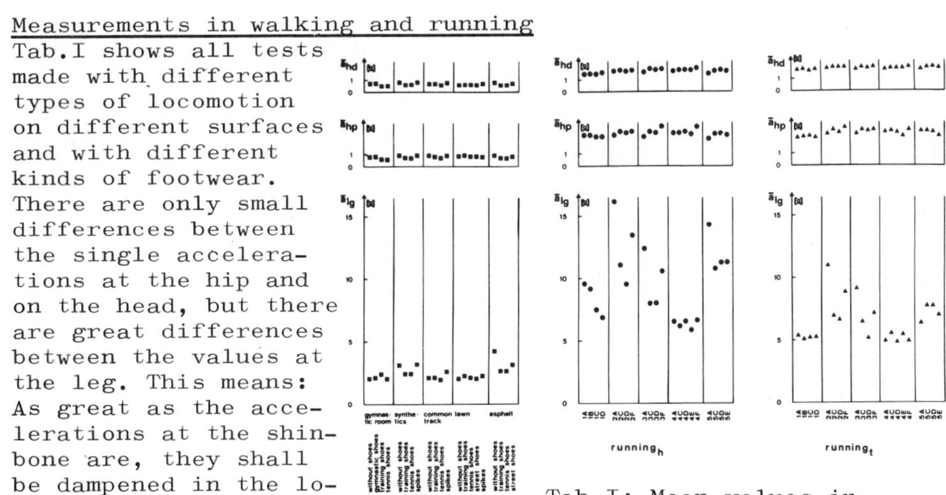

Tab.I: Mean-values in walking and running

Measurements in skiing

Fig.2 shows that in skiing the accelerations at the leg increase quadratically to the velocity, at the hip they increase linearly. This means: in skiing too, the dampening of the accelerations is primarily in the lower extremities. Other tests compared different ski types. Hard and soft skis effect greater values at the leg in downhill than in slalom, soft skis effect greater values in slalom. The surprising thing about this is, that hard skis are commonly used in slalom, soft skis in downhill.

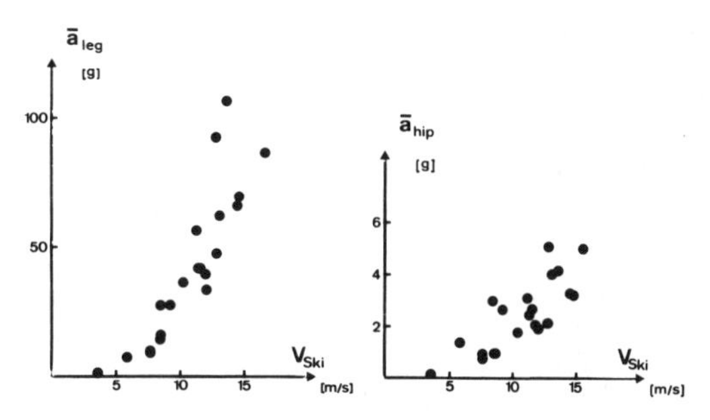

Fig.2: Increase of acceleration to velocity

Measurements in artistic gymnastics

Measurements were made concerning accelerations in different landing types on different surfaces. At the hip the accelerations are 25 % smaller in landing in a deep squatting position or in competition-landing with a roll afterwards than in normal competition-landing. Therefore also the spine is less stressed with these two landing types. In landing on different surfaces great differences appear using the pit. The accelerations at the leg are about 75 % less than on one or two mats. The accelerations at the hip are about 5o % less. In gymnastics the spine is rarely stressed. But these stresses can be reduced. All dismounts and vaults should be executed on a pit and should finish in a deep squatting position.

Results

Tab.II shows a comparison of the single tested movements. The accelerations at the hip are in running and skiing not more than 5g, in artistic gymnastics the values at the hip increase to 2og. This means that in gymnastics the spine has a much greater dampening-work to do. The results are: In running and skiing the greatest stress is in the legs. In artistic gymnastics it is in the legs and especially in the spine. The accelerations were strongly influenced by the type of surface, by the footwear and by the style of landing.

Type of movement	Explanation of the movement	Order of magnitude of the acceleration mean-values		
		a_{leg} [g]	a_{hip} [g]	a_{head} [g]
walking	tennisshoes in gymnastic room	1- 3	o- 2	o-1
	tennisshoes on asphalt	2- 3	o- 2	o-1
	without shoes on asphalt	3- 5	o- 2	o-1
running (heel)	tennisshoes in gymnastic room	5- 8	2- 3	1-2
	tennisshoes on asphalt	1o-15	2- 3	1-2
	without shoes on asphalt	12-16	2- 3	1-2
	with spikes on synthetics	11-16	2- 3	1-2
running (toes)	tennisshoes in gymnastic room	4- 7	2- 3	1-2
	tennisshoes on asphalt	6-1o	2- 3	1-2
	without shoes on asphalt	5- 8	2- 3	1-2
	with spikes on synthetics	7-12	2- 4	1-3
skiing	slope (velocity 7o km/h)	5o-9o	3- 5	2-4
	slope (velocity 3o km/h)	3o-6o	1- 3	1-2
	deep snow (velocity 3o km/h)	4- 6	1- 3	1-2
artistic gymnastics	landing in deep squatting on: 1 mat	18-33	6-12	1-6
	2 mats	16-36	5-13	1-4
	pit	6- 9	3- 6	o-3
	competition-landing:			
	handspring forward (1 mat)	21-28	8-11	2-4
	handspring forward (horse vault, 2 mats)	34-42	8-21	2-7
	salto backward tucked (horizontal bar, 2 mats)	32-59	15-21	4-7
	salto backward tucked (horizontal bar, pit)	1o-15	7-1o	1-4

Tab.II: Comparison of the single movements

NEUKOMM, P.A.; NIGG, B.M. Telemetry: Investigations on Ski Research; IV.Int.Symp. on Biomechanic; Pennsylvania (1973).
NIGG, B.M. Miomechanik, Ausgewählte Kapitel; Vorlesungsmanuskript (ETH Zürich 1973).
SPIRIG, J. Erschütterungsmessungen bei Absprüngen und Landungen im Kunstturnen; Diplomarbeit am Lab.f.Biomechanik (ETH Zürich 1974).
UNOLD, E.M. Ueber den Einfluss verschiedener Unterlagen und Schuhwerke auf die Beschleunigungen am menschlichen Körper; Diplomarbeit am Lab.f.Biomechanik (ETH Zürich 1973).

Biotelemetry II. 2nd Int. Symp., Davos 1974, pp. 115–117 (Karger, Basel 1974)

Precision PCM Telemetry in Ski Injury Research

M.L. Hull and C.D. Mote

Department of Mechanical Engineering, University of California, Berkeley, Calif.

Any skiing injury study must focus on the excitation of the process as it occurs under normal skiing conditions. This paper discusses part of an instrumentation system designed to record the excitation of the boot by the ski. The principle instrumentation problem arises because the excitation is broad band random with a large dynamic range. The low level, low frequency signals and the high level, high frequency signals are equally important. It is also unclear which boot excitation components, or combination of excitation components, are most significant over the range of skiing environments. This necessitates a multichannel system. Portability, durability, and thermal stability are also constraints. The instrumentation topic areas include: (1) the ski, (2) the dynamometers, (3) the signal conditioners, (4) the PCM-FM system, (5) the data processing system. Our attention is directed to signal conditioning and the PCM-FM system, for they are considered to be of general interest.

A PCM-FM system with a 12 bit word length was designed to meet our specifications. It insures minimal transmission error, and a resolution equal to that of the A-D converter. The 100 kHz bit rate and time division multiplexing combine to give 260 Hz resolution on 12 dynamometer plus 1 velocity data channels. The instrumentation system was designed for a 0.1% FS error so that 1% accuracy data is a reasonable expectation. Other systems were evaluated prior to the PCM development. For example, an FM-FM, being essentially a 1% FS system, is not suitable for high frequency, high accuracy data to be reduced by digital computer. A multichannel, miniature, instrumentation tape

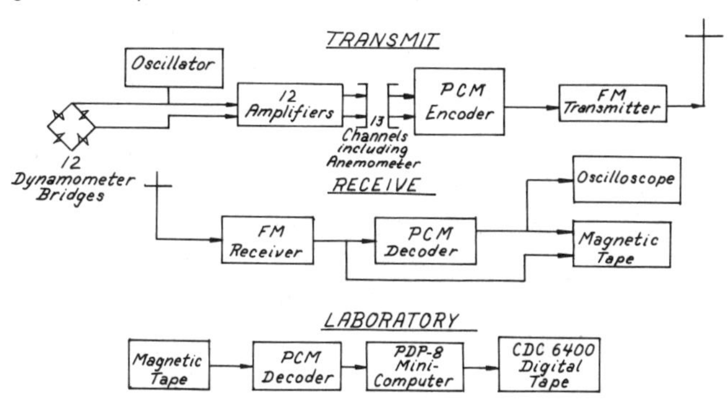

Fig. 1: Complete Data Acquisition System.

recorder of sufficient durability has not come to our attention.

The complete data acquisition system shown in Fig. 1 consists of two six degree of freedom dynamometers and an anemometer for skier velocity. Dynamometer signals are AC-amplified to ± 10V and PCM encoded with the anemometer output. The FM carrier transmits the serial encoder output to the ground station where it is recorded on analog tape. Tapes are decoded in the laboratory and input to a DEC PDP-8 mini-computer, which buffers and formats data onto CDC-6400 compatible tape.

The signal conditioning amplifiers between the dynamometer and PCM encoder present special problems because the stiff, high frequency (300 Hz) dynamometer provides low signals which must be boosted to ± 10V. To preserve the superior signal/noise of the PCM, a carrier amplifier system was designed instead of DC amplifiers. The AC system is preferred for low level, high gain conditioning because it is immune to RF induced voltages. Also, the transducer sensitivity is limited by the strain gage power dissipation capability. For a set power dissipation, an AC system has 183% more sensitivity than a DC system.

A Wein Bridge oscillator generates a 7.5V p-p carrier sine wave of 4.25 kHz and harmonic distortion of about .02%. Ultra stable components were selected for the bridge to provide frequency stability of .05% over the design temperature range, 0°C to 10°C. Amplitude stability, necessary for constant dynamometer sensitivity, is provided by an analog feedback loop which controls amplifier gain.

The three-stage carrier amplifier provides high gain amplification and demodulation of the amplitude modulated carrier. In the first stage, the bridge output is differentially AC amplified at a gain of 200 by an operational amplifier. The AC amplifier output drives the second stage, a passive, phase sensitive demodulator. The final stage, a 10 X DC Op Amp, combines as a second order low pass filter. Tuned for a Q of 0.7,

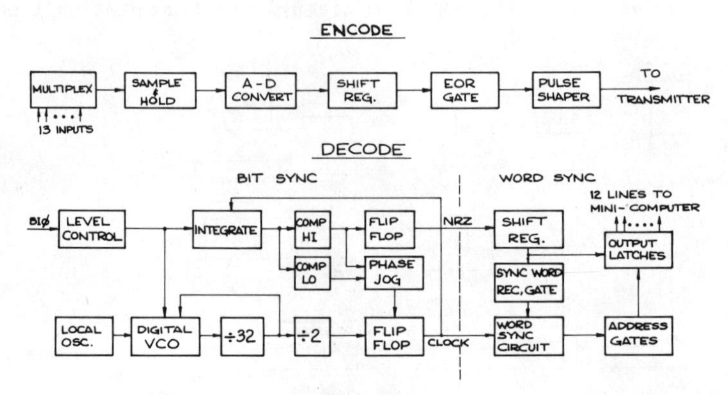

Fig. 2: PCM Encode and Decode.

the filter has flat response to 100 Hz. The maximum DC amplifier swing is ± 10V. For an AC gain of 200 and DC gain of 10, the output noise is 5mV p-p yielding a maximum signal/noise ratio of 2000 to 1. Overall frequency response is 3 db down at 280 Hz.

After conditioning to ± 10V, the 13 analog signals are routed to the PCM encoder, converted to digital form, and prepared for FM transmission. Fig. 2 shows the complete PCM system. The signals feed a 16 channel time division multiplexer which serially switches through each input 521 times/sec, for a 260 Hz theoretical frequency response. The multiplexed values are pulse coded into 12 bit words by an A-D converter. Each 16 word frame contains one sync word, two frame counter words, and the 13 data words. Converter output is further coded into biphase by an EOR gate prior to transmission. The converter uses 2's complement code giving an 11 bit amplitude plus a sign bit. Because inputs are digitized into 2048 levels and converter uncertainty is ± 1 LSB, the maximum signal /noise ratio is 2048 to 1. The highest resolution of the entire amplification and coding process is .1%.

Data, received and recorded on analog tape, is decoded in the laboratory. Decoding is a two-part process, bit synchronization and word synchronization. The bit synchronizer generates a phase locked local clock and decodes the biphase into serial binary bits. Incoming biphase provides the phase reference to a digital VCO which synchronizes the local clock by monitoring the phase difference between the biphase and the local clock. Half cycles from 32 X the local clock are either added or subtracted as required to produce synchronization.

The bit decoder relies on the phase locked clock to synchronize its operation to the incoming data. The biphase is integrated for each half clock cycle and the result goes to a comparator which makes the decision of logical one or zero.

The reconstructed NRZ data and the synchronized clock serve as inputs to the word synchronizer. Here serial PCM is blocked correctly into the original words and word frames. From the shift register, data is parallel output to a sync word recognition gate and seven latches. The gate pulses the word sync recognition circuit whenever any word that has the same binary structure as the sync word is recognized. If four consecutive sync words appear on time (every 16th word), the circuit is synchronized and address gates have access to the output latches. Once synchronized, only the absence of four properly timed sync words can trigger desynchronization.

Acknowledgements: We are grateful to the National Science Foundation for financial support. We also appreciate the assistance of Mr. Ray Mortvedt of the AMF-Head Ski Corp., who provided a specially designed ski. Finally, we thank Prof. Forrest S. Mozer and Mr. Henry Heetdirks of the Space Sciences Laboratory, University of California, Berkeley, for designing the PCM encoder and decoder, and further general advice with this program.

Biotelemetry II. 2nd Int. Symp., Davos 1974, pp. 118–120 (Karger, Basel 1974)

Radio-Telemetric Method of Evaluating Force-Dynamic Values in Alpine Skiing

Georgy Voroshkin

Higher Institute of Physical Culture 'G. Dimitrov', Sofia

Modern Alpine skiing is characterized by too complex and various techniques whose mastering and perfecting is impossible without highly developed force-dynamic qualities.

Building of such qualities is based on characteristics of skier's motion technique in competitive conditions. That makes necessary their studying and evaluating in natural conditions. When using the kinematic method we can obtain only a spatial idea about the skier's motions, which is insufficient for achieving the most correct and effective way of training. Having in mind all that, our purpose was to work out apparatuses and method of research on velocity and force characteristics of technical motion elements. We chose the telemetric method as most reliable and appropriate to Alpine skiing and we constructed a two-channel radio-telemetric system on the basis of accelerometry.

Our study was intended to examine experimentally:

1. Imitative Summer preparatory exercises for the skier.
2. Single elements of skier's technical motions performed in a free manner in natural conditions.
3. A combination of various motions in crossing slalom gates with different positions.

Since the construction and the operation of a similar radio-telemetric system are well known, we show here only its block scheme. (see Fig.1)

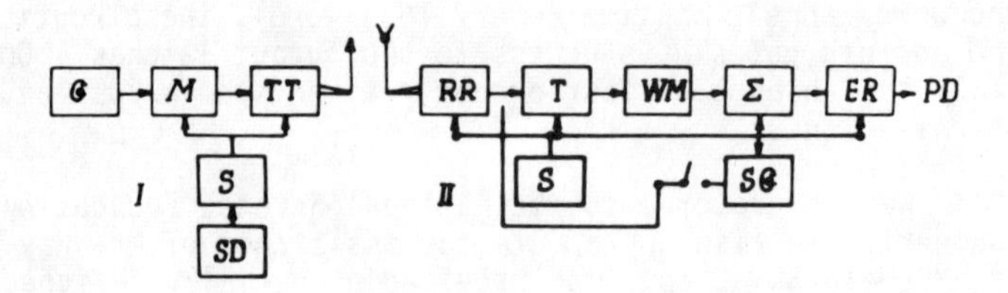

Fig.1
Block scheme of radio-telemetric system

It consists of:(G)-gauge/inertial type/, (M)-modulator/volt-frequency generator/, (TT)-three-step transmitter, (SD)-supply device for accumulators, (RR)-receiver with ultrashortwave range, (T)-Schmitt-trigger, (WM)-biased multivibrator, (Σ)-integrator, (ER)-broadcasting repeater, (PD)-pen-device, (SG)-generator and (S)-supply.
A very important moment in the operation of the measuring channels is their calibration in acceleration of gravity (g).

As the accelerations acting along the line of the body are most significant for creating the efforts in skiing, we use them as a main criterion in characterizing the elements. In order to deduce force values in kgs, we used the already familiar way, namely: Force, exerted by the skier-competitor at a given moment of his performance, can be found from the formula: $F = m.a + P$, where (m) is the mass of the system skier and his equipment, (P)-its weight and (a)-acceleration.

On the tape of the pen-device the acceleration (a) is recorded in units acceleration of gravity (g). From here : $a = y.g$, where y is units of g. Weight $P = m.g$ and $m = \frac{P}{g}$.
By substitution in the main formula we receive:

$$F = \frac{P}{g} \cdot y \cdot g + P \text{ and after elimination: } F = P \cdot y + P.$$

After solving the last equation we can obtain the force values (F) in kgs.

Our first experiment was carried out on 5 imitative Summer preparatory exercises. The subjects in it were five skiers. Accelerograms were made for each exercise and every competitor.

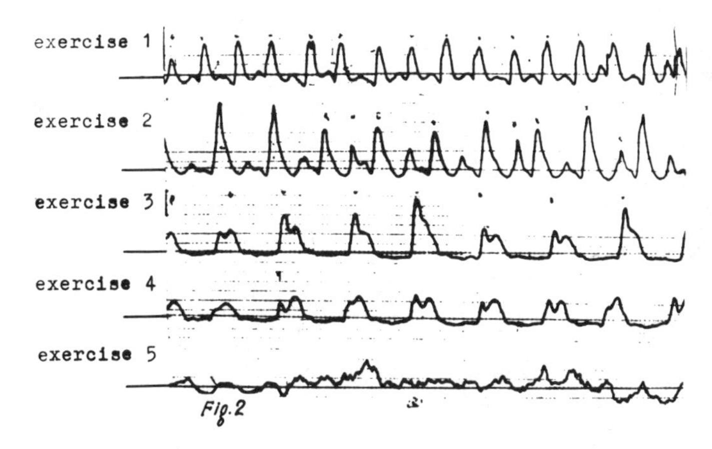

Fig.2 Original record of moment accelerations of the subject N.Sht. for each exercise

DESCENTS	NUMBER OF REALIZED BREAKINGS	MEAN VALUES OF ACCELERATIONS IN g	AVERAGE WEIGHT OF COMPETITORS	MEAN VALUES OF EFFORTS IN KG
1	150	2.25	78	253.5
2	100	1.89	78	225.4

Table I Mean force values for descents 1. and 2.

We show here an original record of accelerations received for 5 exercises of the subject N.Sht.. After reading the accelerogram in the way indicated above, we find minimal, maximal and mean force values in kgs. The mean values for all 5 subjects were: 169kgs, 212kgs, 155kgs, 154kgs and 122kgs respectively.

It is necessary to add also, that each exercise has been performed 12-15 times by every subject and that the indicated values have been drawn out from about 60-75 trials.

The second experiment was in natural conditions on 5 subjects, who realized 20 descents as follows:
1. skiing downhill at little turns (breakings)/10 times/ and
2. skiing downhill at medium turns with gliding steps. The results are shown on Table I. It is evident from it that the average acceleration at middle turns with gliding steps is smaller than that at little rhythmical turns. This difference has an effect on the average force /in kgs/ exerted by the skier at each turn.(see the last column)

The third experiment is crossing slalom gates with different positions-horizontal with great deviations, vertical in direct line, double vertical and horizontal with mean deviations - all 20. 5 subjects were studied at performing 10 descents. The mean data are: 200 gates crossed, mean acceleration in (g):1.97 at average body weight 78kgs and mean force values implied at each turn 231.6kgs .

All that is reported up to here is information about the first attempt in our research work, which is intended to study completely the force characteristics of the technical motion elements in Alpine skiing.

Comment of the editor: Force kgs = kp (kilopond)

Biotelemetry II. 2nd Int. Symp., Davos 1974, pp. 121–123 (Karger, Basel 1974)

Applied Biotelemetry in Skiing

C.-M. Grimm, H. Krexa and E. Asang
Chirurgische Klinik der TU München, München

In order to prevent typical alpine skiing injuries one needs a biomechanical analysis of all the motions employed in uphill and downhill skiing.

Based on the research of skin electromyography and dynamography we needed to obtain certain characteristics and methods to analyse our findings. Two parameters of muscle-physiology define the recorded diagrams of the integral action-potentials:

- the amplitude, representing the number of the stimulated motor-units;
- the frequency, representing the incoming neural impulses.

To analyse the whole sequence of motions we employed a classifiing device. In using this method we classified the integral action-potentials from the diagram with respect to amplitude and frequency; this enabled us to draw a cumulative frequency distribution (fig.1).

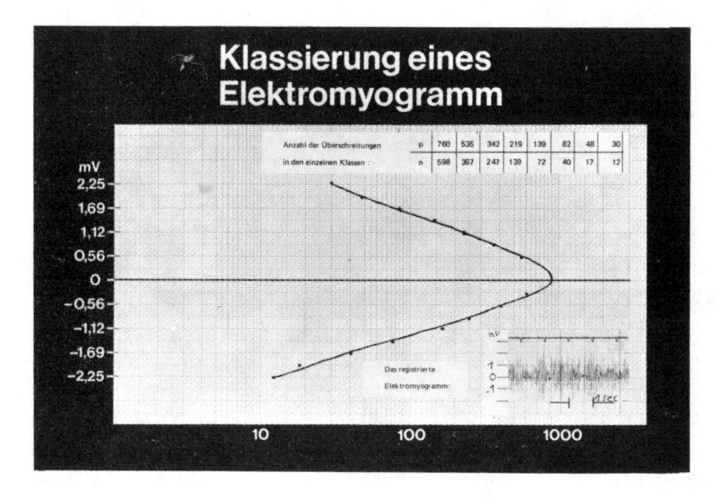

Fig.1 An original electromyogram and its cumulative frequency distribution.

Thus we obtained statistically significant results on the average frequency of the stimulated motor-units and their amplitudes' characteristics (expressed by the angle of the cumulative frequency distribution).

For example the results obtained at the adductors muscle group showed both increasing amplitudes and frequencies going from the herring-bone step to the parallel swing. This evidently is due to the fact that the latter style requires keeping the skis close together.

On the other hand the quadriceps femoris muscle showed similar amplitudes' characteristics for snow-plow turn and parallel swing. However the snow-plow turn demands a higher muscle activity, as can be seen from the 25 percent increase in frequency.

The integral action-potentials determined at the triceps surae muscle predominantly are caused by isometric contractions. This is due to the tensile strain occuring at this muscle during skiing.

The dynamograms were analysed like the electromyogram: either by statistical evaluation with a classifiing device or by individual evaluation of slow-motion plotting. This allows differentiation of steering forces and disturbing forces among the forces occuring at the binding.

Theses forces are of basic importance for determining a reliable retention of the release binding while skiing.

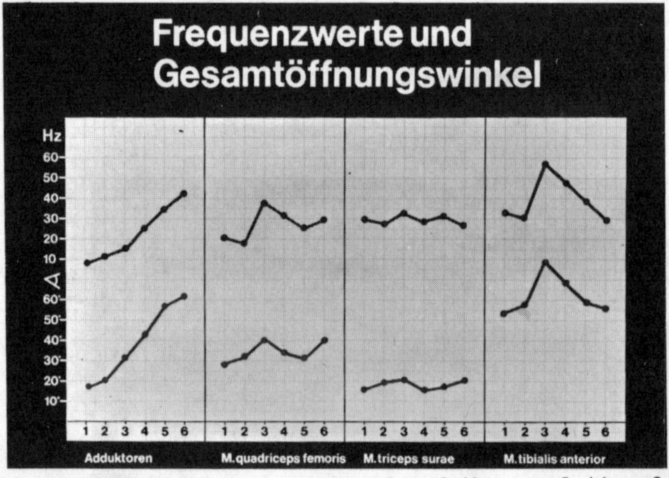

Fig.2 Average frequency and angle of the cumulative frequency distribution in different styles of skiing.
1= herring-bone step; 2= side step; 3= snow-plow turn; 4= stem turn; 5= stem swing; 6= parallel swing.

Our analysing method enabled us for the first time to draw representative cumulative frequency distributions of electromyograms and dynamograms. Results of these curves depend on the individual tested, the muscle group and the style of skiing (see figure 2 and 3).

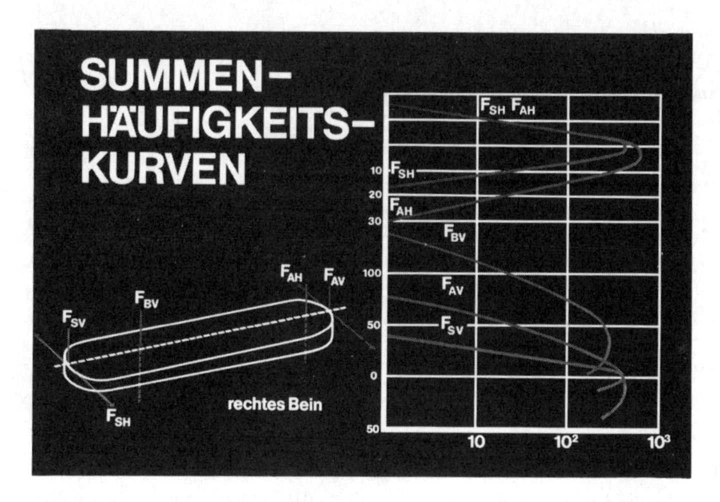

Fig.3 Cumulative frequency distributions of the forces at the binding. F_{SH}, F_{SV} = horizontal and vertical forces at the toe;
F_{AH}, F_{AV} = horizontal and vertical forces at the heal.

To prove the reliability and validity of these analysing methods we transmitted and recorded synchronically electromyographic and dynamographic data. Thus we found the relationship between the activity of the single muscle group and the corresponding forces at the binding. Depending on laboratory characteristics of electromyographic and dynamographic data of a muscle we could use this type of analysis even for quantitative findings.

Hence the interference between muscle activity and the resulting forces is based on a specific scale of measurement even in dynamic sequences of movements.

The new method developed for the registration and evaluation of skin electromyography may be applied to other kinds of sport. It could be of special interest for those disciplines not allowing direct dynamographic control of training conditions and results.

Biotelemetry II. 2nd Int. Symp., Davos 1974, pp. 124–126 (Karger, Basel 1974)

Experimental Biotelemetry in Alpine Skiing

H. Krexa, C. Grimm and E. Asang
Technische Universität München, München

Biomechanical research for the prevention of alpine skiing injuries requires an analysis of the sequences of motions uphill and downhill. We developed a method, which allowed us to relate the activities of leg muscles to the forces in skiing. These specific signals are obtained, transmitted and recorded during field tests by a telemetric set.

We used the electromyography to record the activities of the most important muscle groups of the leg in skiing. The commonly used needle electrodes did not fit for our purpose, therefore we based our research on surface electrodes. The summation of action potentials during a muscle's contraction was obtained superficially. In laboratory studies we established favorable positions of these muscles: quadriceps femoris, adductors, tibialis anterior and triceps surae (Fig. 1).

Depending on the average of the highest amplitudes we have chosen electrodes' favorable positions for the skin electromyogram of defined muscle contractions. The average height of the spikes was in the range of 1,7mV (adductor muscles) and 2,9mV (tibialis anterior muscle). The laboratory results of the electromyographical activity of individual muscle groups can be compared with the measurements found in alpine skiing.

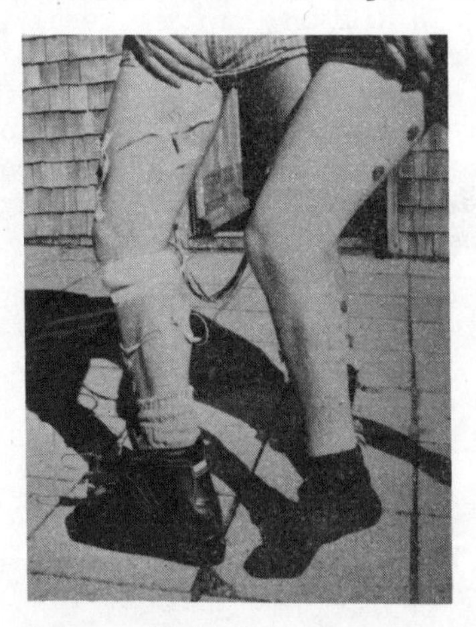

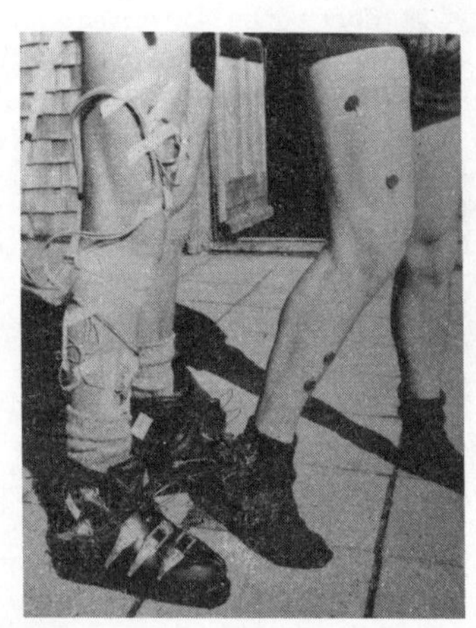

Fig. 1 Positions of electrodes for the electromyograms

The forces generated by muscles result in a specific joint movement. Forces in alpine skiing appear in this chain of events: leg-boot-binding-ski. There were five special impulse receptors put between ski and binding (Fig.2).

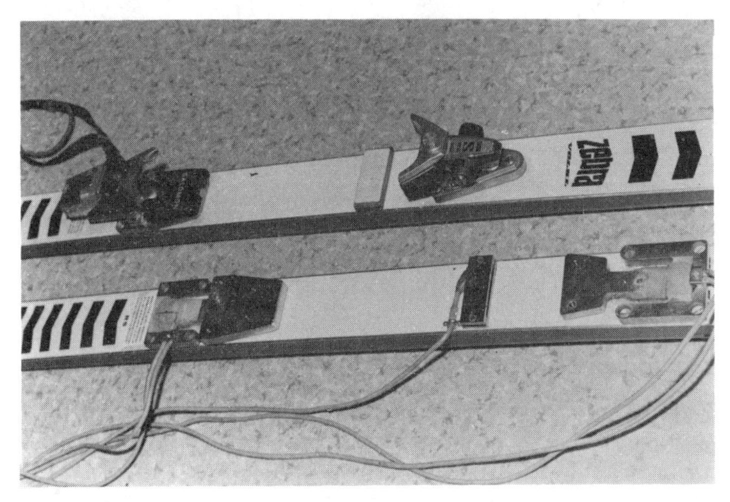

Fig.2 Mechanical impulse receptors for the dynamograms

During skiing all forces were taken by these receptors at the boot's tip and heel and transformed into electrical impulses. Thus all forces influencing the skier's leg can be obtained and transmitted to a telemetric receiving station.

The used telemetric set enables to transmit electrical signals on 4 (or 8) channels without interference and to reproduce electromyograms and dynamograms simultaneously. The muscles' electrical activity can thus be related to the corresponding steering forces developed by the leg.

Figure 3 shows three "Stemmbögen" in alpine skiing from the view of the electromyogram (triceps surae and tibialis anterior muscle) and the dynamogram (horizontal and vertical forces obtained at the boot's tip). The amplitude is gauged in kiloponds (dynamogram) and milli Volts (electromyogram).

Studying the dynamogram you can differentiate the forces into steering forces and disturbing forces. The steering forces lie in low frequency range up to 1/10sec and depend on the individual way of skiing, on the snow temperature and consistency and on the ground relief. The disturbing forces lie in a high frequency

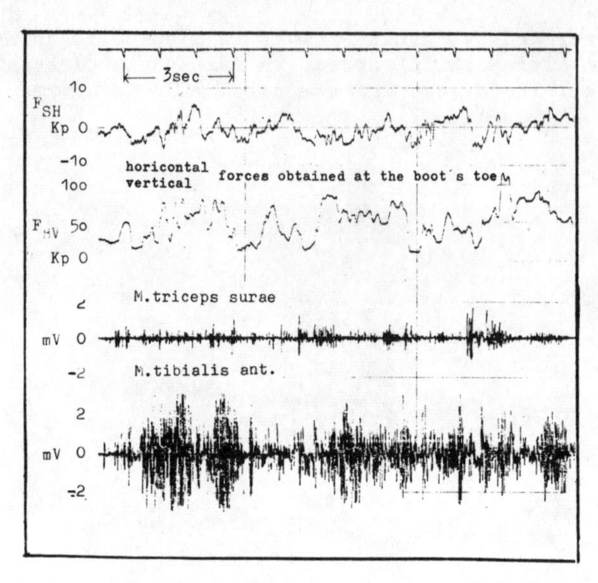

Fig.3 Electromyogram and dynamogram in alpine skiing

range of 1/100 sec and add up to the steering forces. They are primarily caused by the consistency of the ground. They are, of course, high when skiing on an icy slope and with high speed.

The dynamogram may be read from the view of skiing injuries and safety bindings, too. Skiing forces, obtained by this method, are lying below the lowest limit assumed for the release of safety bindings. The rotation forces, for example, ammount to 15kp at the tip and to 30kp at the heel.

The amplitude of the <u>electromyogram</u> - representing the number of active muscular motor units - consists of a great deal of single datas. The electronic way - of dividing the amplitudes into classes of milli Volts and counting the spikes within one class - shows an easy method of reading the electromyography. The investigated muscle is characterised by that frequency of sums. For the first time such curves could be shown in respect to the different test-person, to the muscle and to the individual way of skiing.

The statistical analysis and interpretation of the dynamogram and electromyogram gives a relationship between the activity of a muscle and the extent of the corresponding force. There is still no general relationship between amplitudes, frequency and quantitative force of muscle, but this is a method to find it.

Session VI

Telemetry of Respiratory and Cardiovascular Parameters

Chairmen: *H. Hutten, T.B. Fryer, H. Howald and H.P. Kimmich*

Biotelemetry II. 2nd Int. Symp., Davos 1974, pp. 128–133 (Karger, Basel 1974)

Physiological and Clinical Aspects of the Telemetric Measurement of Respiratory and Cardiovascular Parameters

H. Hutten

Dept. of Physiology, University of Mainz, Mainz

Biological systems are so-called open systems. They exchange energy with their environment. From all kinds of energy, animal cells utilize the energy from only one source, namely from the chemical reactions of metabolism. This energy is used:

a. for biosynthesis, i. e. for the production of specific molecules and the continuous reproduction of cells,
b. for active transport mechanisms, i. e. for amino acids and electrolytes, including absorption and secretion,
c. for all kinds of physical work.

Some of the released energy always appears as heat. Since the body is not a heat engine, it is not capable of utilizing thermal energy. Production of heat means loss of energy. Therefore, a smaller relative heat production signifies a greater efficiency of the system or a more favourable economy, respectively.

The amount of energy which is released within the body, depends on the intake of food and the respiratory uptake of oxygen. Energy stores, such as adenosine-triphosphate and phosphocreatine, are only variables in energy disposition for short intervals. Due to the rather low capacity of the oxygen stores in the human organism - about 3 liters in blood and tissue - and to the restricted availability of energy by anaerobic metabolism which leads to an increase in the blood lactate concentration, the performance capacity is limited to aerobic processes, except in very short periods of exercise, that is for some minutes, and during transitional stages of work.

The intake of food or the maximum absorption capacity of the alimentary canal, respectively, limits the performance capacity only in those cases when the exertion lasts for some hours, days and months.

In all other cases, the performance capacity is defined by the aerobic work capacity and therefore by the oxygen uptake capacity. The amount of oxygen which is necessary for the aerobic processes in the organism is ultimately derived by respiration from the oxygen of the atmosphere.

Respiration means the transport of oxygen from the atmosphere to the cells and, in turn, the transport of carbon dioxide from the cells back to the atmosphere. This process can be divided into four major stages (fig. 1):

RESPIRATION - DETERMINATION OF MAXIMUM OXYGEN UPTAKE		
STAGE	LIMITING PARAMETERS	DISORDERS, DISEASES
1. PULMONARY VENTILATION	MAXIMUM BREATHING CAPACITY	REDUCED COMPLIANCE POLIOMYELITIS ATELECTASIS, PNEUMOTHORAX ASTHMA PARALYSIS OF THE RESPIRATORY MUSCLES INCREASED RESISTANCE OF THE AIRWAYS
	REGULATION OF VENTILATION (NERVOUS, HUMORAL)	ACIDOSIS, ALKALOSIS, WORK, METABOLISM DIABETES MELLITUS
2. ALVEOLAR GAS -BLOOD -EXCHANGE	DIFFUSING CAPACITY	ATELECTASIS, PNEUMOTHORAX PNEUMONIA, PULMONARY EDEMA
	CARDIAC OUTPUT	REDUCED PUMPING CAPABILITY OF THE LEFT HEART DIMINISHED VENOUS RETURN, SHOCK
	OXYGEN CAPACITY AND SATURATION OF BLOOD	ANEMIA, CARBON MONOXIDE, HIGH ALTITUDE LOW INSPIRATORY OXYGEN PARTIAL PRESSURE
	INHOMOGENEITIES OF VENTILATION, DIFFUSION, PERFUSION	AGE, EMPHYSEMA PREGNANCY
3. CIRCULATION	CARDIAC OUTPUT	REDUCED PUMPING CAPABILITY OF THE LEFT HEART DIMINISHED VENOUS RETURN, SHOCK
	BLOOD PRESSURE	HYPOTENSION
	VASCULAR RESISTANCE	OCCLUSION OF VESSELS DURING MUSCLE CONTRACTION THROMBOSIS, STENOSIS, CARBON DIOXIDE
4. BLOOD - TISSUE - EXCHANGE	LOW INTRAVASCULAR OXYGEN PARTIAL PRESSURE	ARTERIAL OR VENOUS HYPOXIA

Fig. 1: Oxygen transfer from atmosphere to tissue cells.

a. pulmonary ventilation, which means the actual inflow and outflow of air between the atmosphere and the alveoli in the lungs,
b. exchange of oxygen and carbon dioxide across the respiratory membrane between the alveoli and the blood by diffusion,
c. convective transport of oxygen and carbon dioxide from the lungs to the capillaries of body tissue and back again to the lungs,
d. exchange of oxygen and carbon dioxide between the blood within the capillaries and the cells of the surrounding tissue by diffusion.

An individual's maximum oxygen uptake is reached when the body is unable to increase further the rate of oxygen supply to tissue. There is a close correlation between the degree of fitness, as referred to physical activity and exercise, and the individual's oxygen capacity. Each parameter which can restrict the maximum oxygen uptake can be a limiting factor for the performance capacity.

The first stage of respiration, the pulmonary ventilation, acts as the controlling quantity in a complex system of regulation circuits. In 1946 GRAY described the ventilation in the famous "multiple-factor-theory" as the variable quantity depending upon the arterial values of oxygen, carbon dioxide and pH. This theory, however, was unable to explain the regulation of ventilation during work. Therefore, in 1950 GRODINS inserted the metabolic rate ratio, which causes an exercise stimulus, into the multiple-factor-theory. Nevertheless, any explanation of the regulation of breathing during exercise remains insufficient up to this date.

In healthy people and referred to normal physiological conditions, however, ventilation is not the limiting factor. The maximum breathing capacity of young people amounts to 180 liters per minute, that is 30 times the value during rest. On the other hand, in patients with severe restrictive or obstructive diseases, i. e. patients with reduced compliance, poliomyelitis, atelectasis or asthma, ventilation can be diminished to a value that is barely enough to maintain life.

The second stage of respiration, the exchange of oxygen and carbon dioxide between the alveoli and blood by diffusion, means the most critical bottleneck. The exchange is controlled by three parameters:

a. pulmonary ventilation, which was already mentioned,
b. diffusion across the respiratory membrane,
c. perfusion of the lungs with blood.

The diffusing capacity of the lungs means the volume of oxygen or carbon dioxide which is transferred within 1 minute and referred to a mean partial pressure difference of 1 mm Hg. For oxygen it is considerably smaller than for carbon dioxide. It depends upon the effective exchange area. Therefore, it is decreased by atelectasis and pneumothorax. However, it depends upon the length of the diffusional path, too. For this reason it is reduced by pneumonia and by pulmonary edema, which may be caused by a reduction in the pumping capability of the left heart.

The perfusion of the lungs nearly equals the cardiac output, because under physiological conditions the pulmonary shunt is a negligible quantity. The cardiac output, which amounts to 5 liters per minute during rest, can be increased to about 25 liters per minute during heavy exercise. Obviously, the increase in cardiac output is considerably less than it is in ventilation. Therefore, under physiological conditions the oxygen uptake is mainly determined by the cardiac output.

Cardiac output by itself is only one factor of a sophisticated carrier system. Firstly, the oxygen capacity of blood must be taken into account. Most of the oxygen is bound chemically to hemoglobin, only a small part of it is dissolved physically. The oxygen dissociation curve of hemoglobin shows a sigmoidal shape. Nearly complete saturation is achieved by normal values of the alveolar oxygen partial pressure. However, the actual saturation also depends on temperature, pH and carbon dioxide, respectively. Incomplete saturation occurs when the inspiratory oxygen partial pressure is low, i. e. in high altitude. On the other hand, the oxygen transport capacity can be reduced either by carbon monoxide or by any form of anemia, i. e. by loss of blood or bone marrow aplasia, etc. Secondly, cardiac output can be diminished, i. e. by aortic stenosis, by insufficient pumping capability of the left heart following coronary occlusion, or by valvular diseases etc. Furthermore, cardiac output can be significantly reduced during circulatory shock. There are people with a predominant tendency to the development of shock, i. e. people with hypotension. A lot of factors promote the tendency, i. e. increased environmental temperature and humidity, dehydration caused by excessive sweating, puberty, circadian rhythms. In such a situation the development of shock can be triggered by an insignificant event, a sudden standing up, by an increase of intrathoraxic pressure and thereby reduced venous return to the heart, as may happen by lifting heavy weights, by defecation or by coughing. In addition, abnormal cardiac rhythms, flutter and fibrillation can be the cause for a reduction in cardiac output. It must be stated, however, that the arterial pressure is merely an auxiliary function for cardiac output, as it depends also on the peripheral resistance.

Thirdly, the exchange efficiency across the respiratory membrane is heavily influenced by any regional inhomogeneity of ventilation, diffusion or perfusion within the lungs.

The third stage of respiration, the convective transport between the lungs and the body tissue, is closely related to cardiac output. The distribution of blood flow to each organ, however, is regulated by different factors. In some organs blood flow is independent on the arterial pressure. In other organs, like the brain, blood flow is controlled by the local carbon dioxide partial pressure. Even emotional reactions are able to change the distribution of cardiac output. During heavy exercise, blood flow to muscle tissue can increase as much as 20 fold. Blood flow to the skin depends upon the environmental temperature. Occlusion of vessels can prevent blood flow to certain organs. Thrombosis is a cause for myocardial infarction. On the other hand, every muscle contraction means a physiological occlusion, during which blood flow is markedly reduced. Therefore, perfusion of the coronary vessels is restricted to diastole. However, an increase in heart frequency during exercise is to the debit of the diastole and therefore, to the coronary perfusion time. As a result, oxygen supply to myocardial tissue is decreased and coronary ischemia can occur.

The fourth stage, the exchange between the capillary blood and the tissue by diffusion, does not mean a critical link in the respiratory chain. Nevertheless, oxygen supply to tissue may become insufficient in cases of arterial or venous hypoxia.

Obviously, the determination of maximum oxygen uptake represents the most valid test for the performance capacity of the respiratory and cardiovascular system. However, as this determination requires that the subjects are exercised extremely violently in order to obtain maximum values, it brings a risk of stressing elderly or diseased people to such an extent. This limits considerably the use of this testing procedure. Furthermore, it has been demonstrated that the peak oxygen uptake depends on the type of work which is performed. As was pointed out, the maximum performance capacity depends significantly on environmental factors, the simulation of which is impossible in surgery. It must be stated that the fitness of a mine worker, of a wood worker, of a man working with a compressed air drill or at a high-temperature furnace cannot adequately be tested on the treadmill.

Oxygen capacity is also of interest with regard to the biological influence of aging on various functions. However, the deviation of oxygen capacity in correlation to age prevents realistic estimations. It must be stressed that a subject's oxygen capacity should always be related to his profession or occupation in order to avert disease from him. Nearly 50 percent of all deceases in the Bundesrepublik are caused by a failure of the cardio-respiratory system. This statement should be sufficient to demonstrate the importance of the telemetric measurement of respiratory and cardiovascular parameters.

However, telemetry cannot only help to prevent failures, it can also help to save a subject's life, if a failure has already occured. The time that is necessary for transporting an accident victim to the hospital means lost time with regard to the emergency treatment. When the patient arrives at the hospital he is still an unknown subject to the medical doctor in this hospital. In such cases a data transfer from the mobile ambulance into the hospital may help to save time. If everybody is equipped with an international health certificate that contains the subject's blood group, the premedication and a brief anamnesis, these data can be transmitted to the hospital by voice communication. In addition, it would be useful to transmit some standard ECG recordings, the arterial pulse shape, some informations concerning the acid-base balance, the electrolytes etc.

Artificial organs will compete with organ transplantations. In both cases, special test methods must be developed. Telemetry may be an adequate solution for some of these problems. In the Bundesrepublik about 20 000 subjects have an implanted artificial heart pace maker. However, there is always an important factor of insecurity, namely the energy source of the pace maker. For this reason, these people have to visit the hospital every two months. In the meantime they are exposed to the fear that a failure of the battery may occur. Proposals have been made to use the normal telephone system for the transmission of ECG and for the examination of battery charge. Another point is that in some patients even artificial pace makers cannot prevent disorders of the normal rhythm. It would be very important in those cases to have a direct link connecting the patient to the hospital. Therefore, an international agreement should be achieved for biotelemetry via telephone system.

In many cases telemetric measurements of respiratory and cardiovascular parameters can be performed by applying noninvasive methods, i. e. transcutaneous measurement, by means of ultrasound or thermal methods etc.

References:

1. GRAY, I. S.: The multiple-factor-theory of the control of respiratory ventilation. Science 103, 739 - 744 (1946).
2. GRODINS, F. S.: Analysis of factors concerned in regulation of breathing in exercise. Physiol. Rev. 30, 220 - 239 (1950).

Biotelemetry II. 2nd Int. Symp., Davos 1974, pp. 134–136 (Karger, Basel 1974)

Environmental Studies on Blood Pressure, ECG and EEG Using a Programmable Four Channel Telemetry System*

R. Zerzawy, H. Fleischer and K. Bachmann

Med. Poliklinik, University of Erlangen, Erlangen

Economical application of multi channel telemetry systems in environmental studies necessitates free and uncomplicated programmability to various parameters.

Method: A new miniature preamplifier has been developed with programmability to EEG, EMG, ECG, direct intravasal blood pressure measurements, respiration curve, heart sounds, strain gauge bridges and DC- or AC-sources from 10 mV to 1 V (Fig. 1). The function is determined by 12-pin coded input plugs, selecting adequate feedback networks and time constants. The basic circuit consists of an operational amplifier (Q_4), and a differential FET input stage (Q_1, Q_2) with a constant current source (Q_3). EEG, EMG and ECG are amplified 200 to 6000 times, this gain being controlled by R_{10}. Further data reads: time constant 1.5 sec., cmr 80 dB minimum, input noise level below 5 uV rms.

Systemic and pulmonary artery blood pressures are transduced by STATHAM bridges P 23 Db or SP 37, and amplified by Q_4 used as a DC differential amplifier. The pressure range is -30 mmHg to +300 mmHg, at a voltage gain of 180. R_1 provides zero alignment; the long term drift is be-

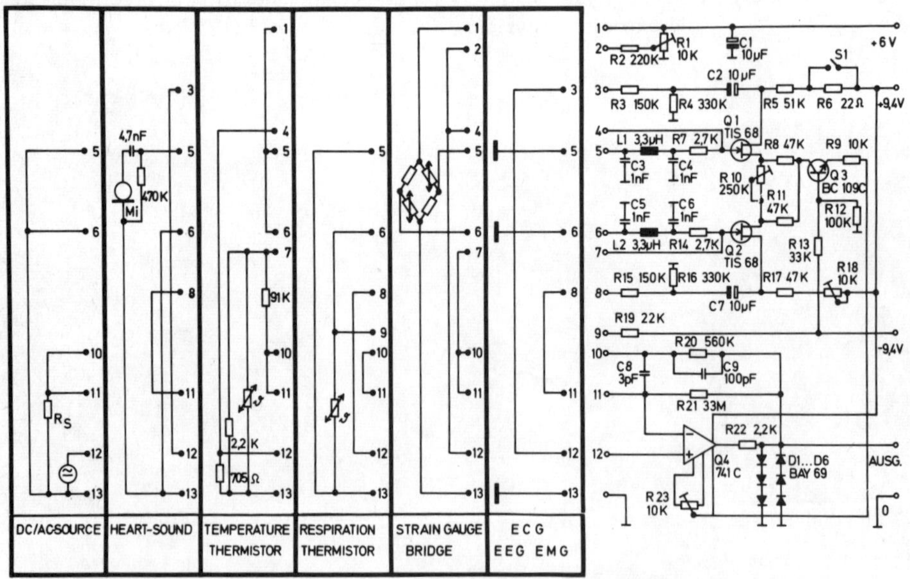

Fig. 1 Circuit diagram of the programmable preamplifier.

*Supported by a grant of the Deutsche Forschungsgemeinschaft

low 1%/hr and the temperature drift less than 0.1 %/°C. The adaption of further transducers and signal sources is shown in fig. 1. This type of preamplifier is used in a four channel telemetry system (3), the transmitter unit weighing 1500 g, having an operation time up to 15 hours.

Combined Radiotelemetry of Direct Arterial Blood Pressure, ECG and EEG during Automobile Driving: Sustained alternations of blood pressure and heart rate as well as transient stress effects (particular traffic situations, meals, intermittent physical activities) were studied in 11 healthy subjects. Information about the daytime factor was given by 12 periods of rest. Each subject drove the same route of 550 km, the EEG being used as an indicator for overtiredness.

Before starting in the morning, mean values while sitting in the car were 130/79, Pm 99 mmHg, heart rate 83/min. (Fig. 2). The greatest relative increase of heart rate to 101/min. was recorded in the first minutes of driving (start reaction), with only moderate increase of blood pressure (139/81, Pm 103 mmHg). During the following half hour pressures slightly decreased to 134/81, Pm 101 mmHg, whereas heart rate showed a distinct decline to about 91/min. After the first period of sport activities there was a marked reduction of resting blood pressure (112/74, Pm 88 mmHg), and an increase of heart rate (105/min.), resulting from postexercise hypovolaemia. The following driving period, including lunch, shows constant mean values of blood pressure and heart rate. A momentary increase of heart rate to 150/min. (blood pressure 160/85 mmHg) in this period was the most extreme response seen in our material, caused by a dangerous condition. The long distance drive in the evening proved a continuous decline of heart rate. Blood pressure values remained remarkably constant, except for a moderate increase to 133/87, Pm 104 mmHg, at supper.

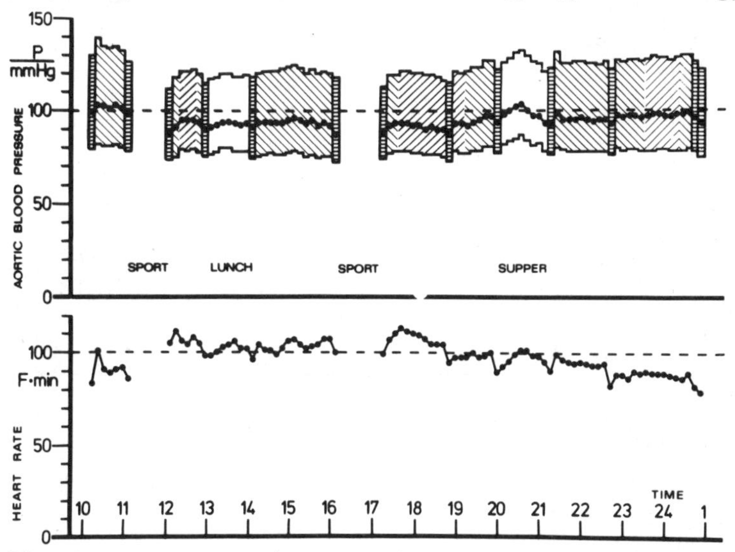

Fig. 2 Results of radiotelemetric measurement of direct arterial blood pressure and heart rate during car driving.

Simultaneous Radiotelemetry of Direct Arterial Blood Pressures During State Board Examinations: In 15 medical candidates (Figs. 3 and 4) the answering of questions provoked hypertensive blood pressure values (145/93, Pm 113 mmHg) and a moderate tachycardia of 97/min. The increase is significant to rest (121/78, Pm 95 mmHg, heart rate 81/min.) and waiting for questions (133/85, Pm 105 mmHg, heart rate 88/min.). During 15 minutes of a written examination blood pressure (135/87, Pm 107 mmHg) and heart rate (91/min.) remained somewhat lower than those of oral examination. Both responses show a remarkable rise of diastolic pressure.

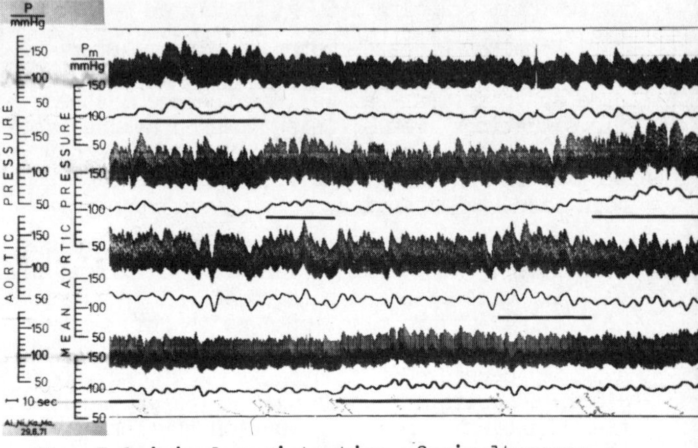

Fig. 3 Original registration of simultaneous telemetry of four arterial pressure curves with integrated mean arterial pressures during oral examinations.

Compared with blood pressure responses to dynamic exercise, mental stress provokes an increase of cardiac output, the peripheral resistance remaining almost unchanged.

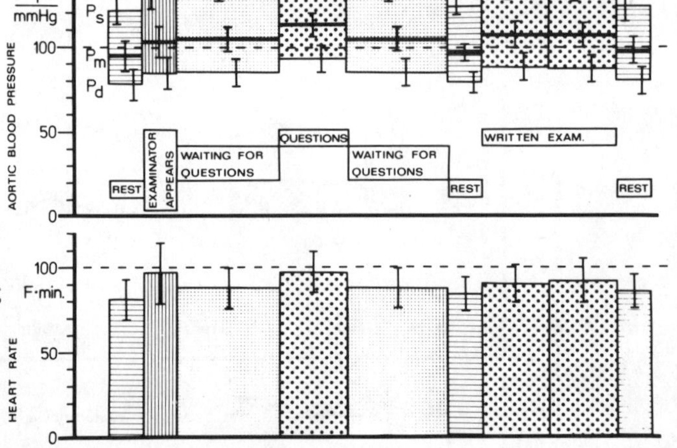

Fig. 4 Results of arterial blood pressure telemetry during the whole procedure of state board examination.

References:

1. H. HOFFMANN und W. REYGERS: Kreislaufuntersuchungen bei Kraftfahrzeugführern unter variierten Fahrbedingungen. Zentralbl. f. Verkehrsmed. 3; 131 (1960).
2. Th. v. UEXKÜLL und E. WICK: Die Situationshypertonie. Archiv f. Kreislaufforsch. 39; 236 - 271 (1962).
3. R. ZERZAWY and K. BACHMANN: A programmable four channel telemetry system for long term radio telemetry of biomedical prameters. In: H. P. KIMMICH, J. A. VOS: Biotelemetry (Leiden 1972).

Biotelemetry II. 2nd Int. Symp., Davos 1974, pp. 137–139 (Karger, Basel 1974)

Combined Telemetry of Cardiovascular Parameters in Sports*

Continuous Measurements of Direct Aortic and Pulmonary Blood Pressures

H. Fleischer, R. Zerzawy and K. Bachmann

Medizinische Poliklinik, University of Erlangen, Erlangen

Biotelemetry as a routine method in sports is up to now almost synnonymous with ECG-telemetry.The continous registration of direct arterial blood pressure(2),is an important step to a more complete analysis of cardiovascular reactions during exercise.Little is known about the hemodynamics and potential cardiovascular hazards of widely spread sport activities such as skiing,rowing,swimming and "fitness trails".

Method:A red ÖDMANN-LEDIN catheter filled with 0.9% heparinized saline solution was inserted into the abdominal aorta by the transfemoral SELDINGER technic and attached to a strain gage(STATHAM P 23 Db or SP 37). In measuring the pulmonary artery pressure a COURNAND F-8 catheter was inserted percutaneously via the femoral vein or a SWAN - GANZ balloon catheter via the basilica vein.The pressure transducer was fixed presternally at the 2nd rib level.The right atrium,being the zero reference level,has about the same distance from this point in the upright,as well as in the supine position.

This method has proven to be safe without restricting the various activities of the subjects in over 1800 investigations.

Skiing: Radiotelemetric studies of direct aortic blood pressure and ECG on 16 healthy students (Fig.1).Walking uphill and cross country skiing(dynamic exercise) were associated with a moderate increase of systolic and almost unchanged diastolic pressures(159/82, Pm 114 mmHg - 158/80,Pm 113 mmHg).In contrast,downhill skiing(static exercise)provoked a significant increase in both systolic and diastolic pressures(185/101,Pm 132 mmHg),heart rate being almost identical in all three exercises(175,172,169 beats/ min).This is consistent with

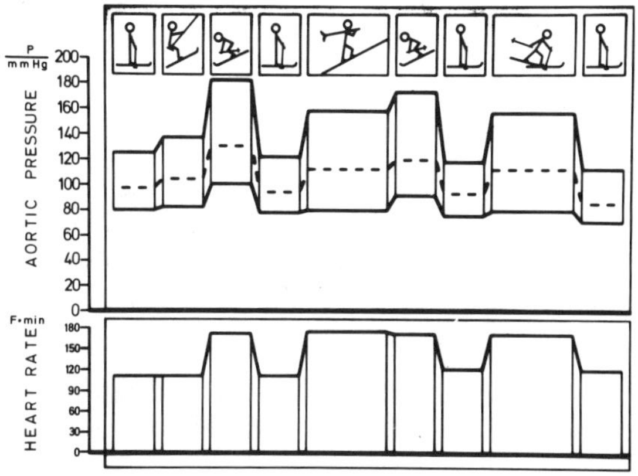

Fig.1.Direct arterial blood pressure, mean arterial pressure and heart rate:mean values of 16 subjects during skiing.

*Supported by a grant of the Deutsche Forschungsgemeinschaft

a prevailing pressure load in the latter and a prevailing volume load in the first type of skiing.

Rowing:Combined radiotelemetry of direct arterial pressure and ECG on 10 healthy students (Fig.2,Fig.3).Continuous rowing for 1 hr resulted in a moderate tachycardia (about 150 beats/min)with moderate increase in systolic and diastolic pressures(155/100,Pm 121mmHg).

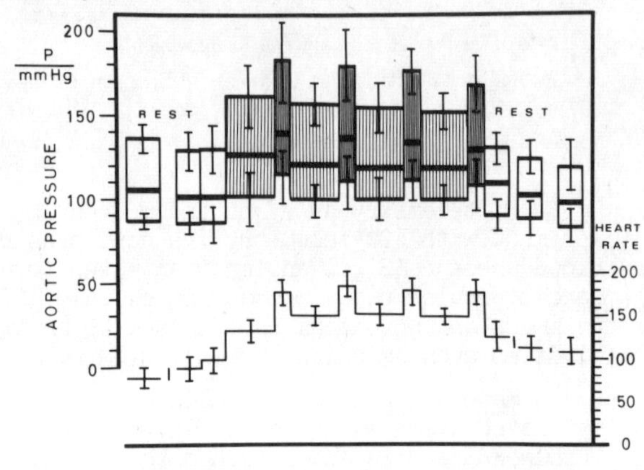

Fig.2.Direct arterial blood pressure, mean arterial pressure and heart rate:mean values of 10 subjects during rowing.

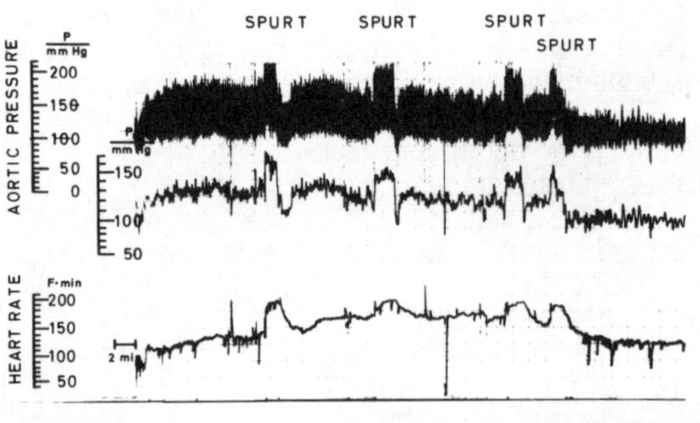

Fig.3.Radiotelemetry of direct arterial pressure with integration of mean arterial pressure and heart rate in a 23 year old student rowing for one hour.

Rowing with maximal strength for two minutes produced a sharp increase in heart rate(180 beats/min),as well as in systolic and diastolic pressures (176/111,Pm 135 mmHg).

Swimming:Combined radiotelemetry of pulmonary artery pressure,aortic pressure and ECG on 16 patients with chronic obstructive lung disease, showed a pulmonary hypertension of moderate severity(Fig.4).

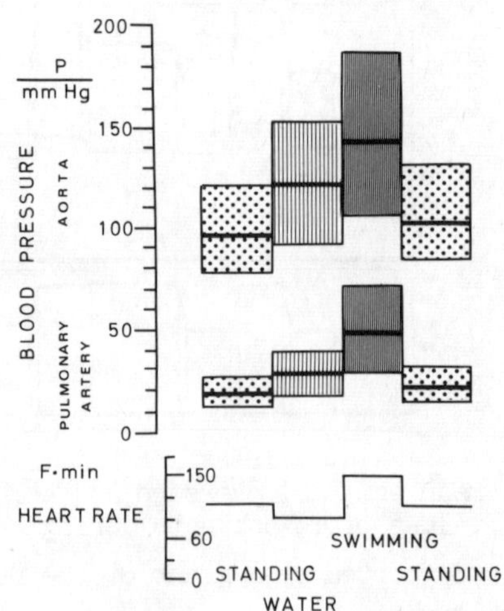

Fig.4 Direct systemic,pulmonary artery,mean arterial pressure and heart rate:mean values of 16 subjects with chronic obstructive lung disease

Results of combined radiotelemetry on "fitness trails": Continuous registration of blood pressures and ECG on 8 healthy students. The volunteers were requested to run between 15 different exercises on a "fitness trail" (total distance 2300 m, h 30m). Fig. 5 shows that already after a few minutes a maximum heart rate averaging 190 beats per minute is reached and lasts until the end of the trail. While the blood pressure, during running, is only slightly elevated (150/80, Pm 110 mmHg) there are marked differences in pressure response, depending on the kind of exercise. Prevailing static work provoked the highest pressures - chin ups (207/130, Pm 152 mmHg), sit ups (189/77, Pm 119 mm Hg) and push ups (187/122, Pm 142 mmHg). Pressures recorded during gymnastic exercises did not differ markedly from those recorded during running.

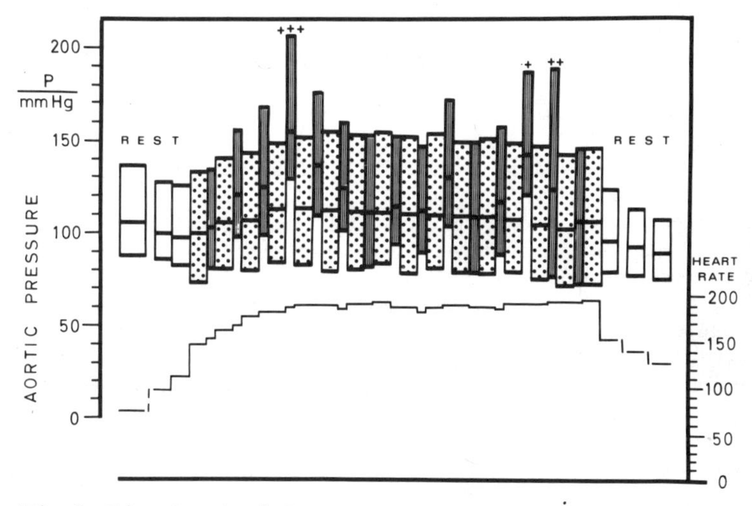

Fig. 5 Direct arterial pressures, mean arterial pressures and heart rate: mean values of 8 subjects during a "fitness trail". +++ chin ups, ++ sit ups, + push ups. Spotted columns - intermittent running.

Conclusion: The results show clearly that there are marked differences in cardiovascular response due to different sport activities, differences which are not apparent with ECG - telemetry alone.

In dynamic exercise, the increase of the work load of the heart is due to an increase of cardiac output with low peripheral resistance but little elevation of blood pressure. In contrast, with static exercise there is a considerable increase in pressure load, which may be unexpectedly dangerous to a person with inapparent heart disease or impaired integrity of the arterial wall (1).

References: 1. DONALD, K.W., A.R. LIND, G.W. McNICOL, P.W. HUMPHREYS, S.H. TAYLOR and H.P. STAUNTON: Cardiovascular responses to sustaind (static) contractions: Circulation Res. 20; suppl. I, 15-30 (1967)

2. ZERZAWY, R., K. BACHMANN and A. HENNING: Simultane drahtlose Telemetrie von mehreren biologischen Messgrössen: Z. Kreislaufforschg. 60: 162-169 (1971)

Biotelemetry II. 2nd Int. Symp., Davos 1974, pp. 140–142 (Karger, Basel 1974)

Radio-Telemetric Study on Heart Rate of Runners

Kliment Boitchev

Higher Institute of Physical Culture 'G. Dimitrov', Sofia

The progress of modern sports requires the systematic control and studying of the athlete's condition. This is due to the fact that nowadays there exists a clear trend to increasing the strenuous work done by sportsmen in comparison with the time even ten years ago, and which will probably continue to increase in the future.

Radio-telemetric studies on the cardiac activity of man is mainly developed in three directions: 1. In the field of cosmos science, 2. In clinical medicine and 3. In the field of sports and labor.

In all cases the intentions are either to investigate the functional reactions of man when working and characterising them, or to regulate those reactions in order to achieve a definite effect from work performance. In the first case, data is recorded and interpreted and thus the applied strain is characterized, and in the second--the data is recorded, interpreted and the values of the athlete's behavior are used for it's correction. The last is mostly applicable in the field of sports.

There are many reports on the above mentioned problems. Among them are the works of V. Rosenblat , A. Vorobiov, M. Kazakov, R. Uzjin, V. Utkin, V. Zatziorski, S. Sarsania, N. Kulik, V. Kalinin, N. Pudov, F. Suslov, N. Holter and his group, V. Seliger, J. Dendal, H. Beenken, T. Winsor, etc.

The purpose of our researches was to examine experimentally the heart rate as an integral functional index in training and by relating it to the parameters-distance (S), velocity (V) and time (T), to find indices characterizing the athlete's behavior at different distances, as well as to control the effect of the applied strain and the athlete's behavior in order to achieve highly effective training. These investigations were intended to help coaches in their work. We used Telemetric System 1C-45 Mk-4 channels.

The subjects in our study were 32 runners--7 sprinters and 25 long and middle, good and excellent runners between 18 and 25 years of age.

Table I. Obtained mean values from sprinters
n-number of trials

n	S	Ttot	T100	Vm/sec	Ptot	P100	Pmin	Ps and Pm
24	100	10"85	10"85	9,22	30,02	30,02	166	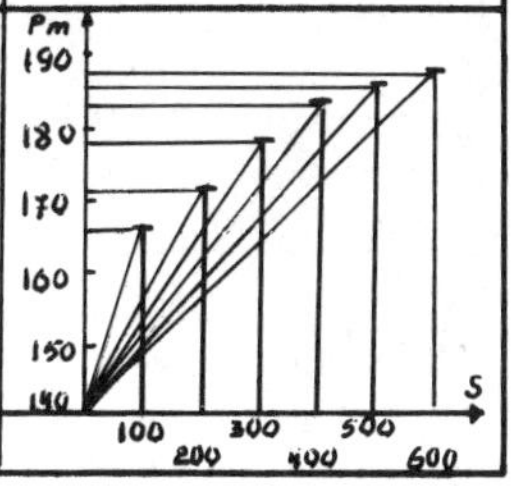
24	200	21"5	10"75	9,30	61,27	30,64	171	
21	300	34"7	11"57	8,64	102,94	34,31	178	
20	400	50"5	12"63	7,92	154,02	38,51	183	
12	500	69"3	13"86	7,22	213,67	42,73	185	
9	600	93"2	15"53	6,44	288,92	48,15	186	

Sprinters performed runs at different distances from 100 to 600 meters at their top speed. Our task was to determined: 1. Total T (Ttot), 2. Mean T for 100m (T100), 3. Mean V for the distance (Vm/sec), 4. Total heart rate (Ptot), 5. Mean heart rate (Pm) starting (Ps); per second, per minute (Pmin) and per each 100m (P100). /see Table I./

Middle and long distance runners performed runs exceeding 6000m at a speed within the range of maximal aerobic work capacity /PWC 170-180/. We determined: 1. Mean T per each 100m, 2. Mean V for the whole distance, 3. Mean heart rate (per each 100m, per minute and per second).

In both cases we were interested in the reactions of the subject's organism and the efficiency in his behavior. For determination of those parameters we used the quantitative ratios among T,V and P.

The data illustrated in Table I. characterize time and velocity parameters in runs at maximum speed at distances from 100 to 600m, heart rate as a relative energy indicator of functional intensity of work. In fact, coaches are interested in the ratio speed/heart rate. The lower the heart rate level is at equal V, or the higher V is at equal heart rate level, running can be considered more effective and the prospect of improved performance more promising. Of course, in practice we took into consideration the individual quantitative specificities of the subjects. An object of consideration was also moments, connected with increasing of speed throughout the distance, increasing of heart rate, the intensity of recovery and many other parameters.

From the studies on long and middle distance runners, we obtained very rich information (see Table II.).

Table II. Summarized mean values from long and middle distance runners.

trials	T100 sec	Vm/sec	P100	Pmin	Psec	Vm/sec / P sec
56	22"23	4,50	65,32	176,3	2,94	1,50

The correlation between heart rate and speed intensity determines the runner's work capacity to perform economically from the energy point of view and effectively from the velocity aspect. The presented quantitative values are deduced at eliminated starting values of P, T and V, i.e. during a run atsteady state around the highest level of aerobic work capacity according to the still-existing ideas in Physiology (PWC 170-180). From practice we drew the conclusion that the levels indicated above are useful for the effective control over the training process mainly during the preparatory period. Later on, during basic, and especially in competitive periods, they demonstrate steady state at heart rates above 180 (see graph in Table I.) and evidently a higher speed. This also required taking into consideration the individual work capacity. During the training of runners we determined some other values of a functional and speed nature. This complex way of research helped both the coaches and us to find the most progressive forms of training. It is well known that Bulgarian track-and-field competitors have had good success in recent years.

The data reported here does not finish the problem. Its deep investigation is an object of our future work.

Conclusions:

1. Researches on T, V, and P parameters during training ensures a rich information helping the correction and the right way of training of runners.
2. Running at PWC 170-180 has a good effect only during the first periods of training. Those levels have to be corrected individually during basic and competitive periods.

Biotelemetry II. 2nd Int. Symp., Davos 1974, pp. 143–145 (Karger, Basel 1974)

Breath Frequency Analysis of Different Sports Athletes During Competition

R. Deroanne, L. Delhez and J.M. Petit
Institut E. Malvoz, Liège

Several authors, particularly DEJOURS et coll (1956) put in evidence the neuro-humoral control of breathing during laboratory tests. This fact is mainly illustrated by the "immediate rise" and the "instantaneous fall" of ventilation respectively at the beginning and at the end of muscular exercises (DEJOURS, 1964).

The aim of this work is to verify the presence of the neuro-humoral theory of breath control during sport practice and especially during competition.

For that purpose we studied the ventilatory behaviour of different types of athletes : sprinters, middle-distances runners, soccer players, swimmers and rowers.

The diaphragm electric activity is derived from an oesophageal catheter (PETIT et coll. 1960). It is then recorded on a 700 gr weight SONY TC 40 recorder which presents a linear frequency range from 30 to 3.000 Hz. In some experimental circumstances (sprinters and rowers) we simultaneously recorded both the diaphragm and the peripheric muscle activities to study the eventual influence of the latter on the former. An example is shown in figure I which represents the diaphragm during a 100 meters run.

This study shows three types of reaction with respect to the sport activity.

1. In a sport which is free of cadency problems and when the evolution environment does not impose a specific adaptation, the respiratory pattern seems to be controlled by three distinct mechanisms the action of which interfere with each other.

a) during a few seconds just before and after the start of a sprint run, the athlete holds his breath to provide a good fixation of his thoraco-abdominal musculature, point d'appui of the arm movements. We may speak then of a voluntary breath control, from a central impulse.

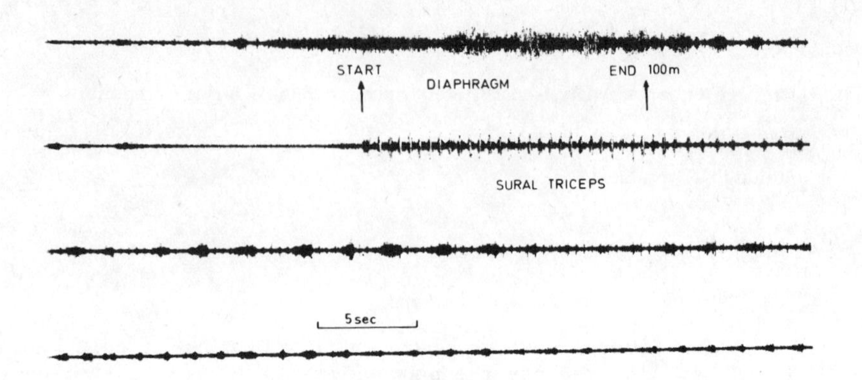

b) after that breath holding, a very quick adaptation of respiration is observed corresponding to a sudden variation in the run speed. These sudden ventilatory rate adaptations confirm the examples described by DEJOURS et coll in 1956, and which seem to be induced by a nervous stimulus from a mechanical.proprioceptive origin.

c) following these quick ventilatory rate variations, slower and reduced ventilatory rate modifications occur when the athlete reaches a relatively stable speed. They may be explained by humoral modifications.

2. When the athlete works in a particular environment (swimming) or when he must perform hard and rhythmed efforts, we observe a ventilation control imposed by the sport technical modalities.

In swimming,two different behaviours are put in evidence, as a function of the style.

a) in breaststroke and crawl the swimmer must be able to breathe out when his head is immersed in order to take the opportunity of the brief head emersion to draw a breath as deep as possible.

b) in backstroke, though the swimmer has his face out of the water, the ventilatory rate is disturbed by the thoraco-abdominal effort due to the arm elevation out of the water. The inexperienced swimmer has then an incomplete expiration.

In rowing two factors lead the ventilatory rate : on the one hand the important thoraco-abdominal effort during the traction on the oars; on the other hand the abdominal viscera compression during the legs- and trunk flexion while the oars are pushed forward. Acting on the diaphragmatic cupola this last movement helps expiration. Both these factors regulate the ventilatory rate of rowers who take breath once or twice during traction phase while expiration is made during recovery phase.

The study of the athlete ventilatory rate confirms the presence of a neuro-humoral control when he freely runs on a sport field.

A cortical influence can induce different breath control when the athlete must either perform important thoraco-abdominal efforts (run starts, oars pulling) or evoluate in a particular environment (swimming).

DEJOURS, P., TEILLAC, A., LABROUSSE, Y. and RAYNAUD, J. Etude du mécanisme de la régulation cardio-ventilatoire au début de l'exercice musculaire. Rev. Franç. Etudes Clin. Biol. 1, 504-517, 1956.

DEJOURS, P. - Control of respiration in muscular exercise. In : "Handbook of Physiology - Respiration, Washington D.C. Am. Physiol. Soc., 1964, sect. 3, 1, 631-648.

PETIT, J.M., MILIC-EMILI, J. and DELHEZ, L.
Examen de l'activité électrique du diaphragme par voie oesophagienne chez l'homme normal.
J. Physiol (Paris) 1960, 52, 190-191.

Biotelemetry II. 2nd Int. Symp., Davos 1974, pp. 146–148 (Karger, Basel 1974)

Respiratory Telemetry in Exercising Horses

H. Hörnicke, H.-J. Ehrlein, G. Tolkmitt, H. Husch, M. Nagel, D. Decker, E. Epple, H.P. Kimmich and F. Kreuzer

Abteilung Zoophysiologie, Universität Hohenheim, Institut für Biomedizinische Technik, Stuttgart, and Dept. of Physiology, University of Nijmegen, Nijmegen

Little is known about the pattern and regulation of respiration during exercise in animals. One might expect that in quadrupeds the forceful action of the legs on the thorax would result in a breathing type considerably different from that in man. Only a breath by breath analysis of respiration in freely moving animals can provide such information.

Method: A multichannel telemetry system is used to investigate relationships between respiration, gait and speed of horses under the rider. Respiratory telemetry is part of an approach to measure oxygen consumption in exercising horses (HÖRNICKE et al. 1974).

Respiratory air flow is measured by a strain gauge pneumotachograph (KIMMICH, VOS & KREUZER 1972) adapted to the respiratory flow range of the horse. The horse wears a face mask airtightly fitted to the animal's head by a rubber maskholder (Fig.1). A flexible connection between this mask holder and the bit allows the horse to be ridden without damaging the rubber. A fiberglass mask with the respiratory tube is fixed to the mask holder by a flange and screws. The air flow sensor consists of a sickle-shaped flag centered in the respiratory tube with a pair of strain gauges at its base. Deflection of the flag is measured by a carrier-frequency amplifier. To compensate

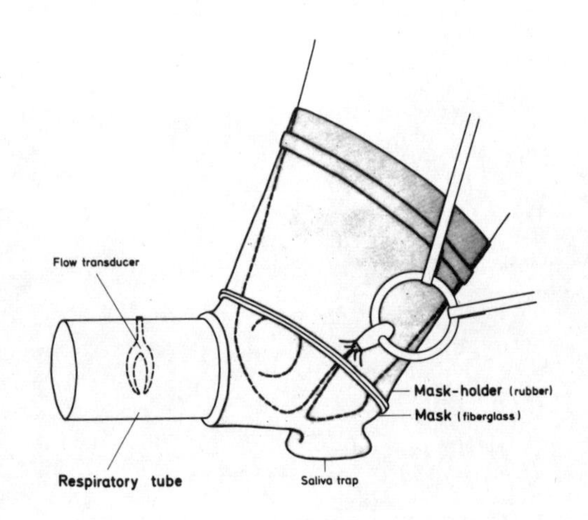

Fig.1 Mask holder, mask and respiratory tube in position.

for effects of gravity and acceleration, the tube contains a second pair of strain gauges at a flag of equal mass but little air resistance. The respiratory tube is calibrated against rotary displacement meters up to 70 l/s at the municipal gas works. The square root of the signal is essentially proportional to air flow. Because of this nonlinear characteristic it is not possible to cover the whole range of flows from rest to heavy exercise with a single tube diameter. At low air flows during rest and pace the sensitivity of the device is therefore increased by an insert reducing the internal diameter of the tube.

Step frequency of the animals is monitored by a strain gauge attached to a foam rubber bandage at a metacarpal joint.

Velocity of the horse is obtained from signals produced by the rider with a hand switch when passing distance markings.

Transmission and storage: The rider carries amplifiers, transmitter and batteries in a vest. The signals for flow, step frequency and distance are transmitted by an FM/FM phase lock loop system (Teldata 500, Glonner Electronic München). A fourth channel is used for tape speed correction. The multiplexed signal is received and stored on one channel of a two-channel magnetic tape recorder.

Data processing in the laboratory: Signals are demodulated and recorded for interpretation. The pneumotachogram is linearized by a four quadrant multiplier (Function Moduler, Irvine, Calif.). Tidal volume (V_T) is obtained by an respiratory integrator. Intervals chosen for computer evaluation are marked by a digital signal on the second channel of the tape recorder.

Computation: The demodulated signal is fed on-line into a PDP 12 computer. It is converted at a rate of 50 per sec. to digital form and appears on the screen. Since horses often have polyphasic inspirations and expirations, special program logic was necessary to recognize and process interrupted half cycles. The program converts the flow signal to l/s STPD. The on-line program stores duration and volume of each inspiration and expiration on magnetic tape. It also prints the number of respiratory cycles, the mean duration of inspiration and expiration (T_I and T_E) and the total volume inspired and expired for the interval given by the command channel. From the values stored on digital tape an off-line program calculates derived values as e.g. respiratory ratio or mean inspiratory flow. The derived as well as the stored parameters may be shown numerically or graphically on the screen or printed out by teletype.

Results: The method was routinely used in the field. Four horses were studied repeatedly under the rider during a standardized sequence of standing, walking, trotting and gallopping at different speeds.

1. Pattern and Synchronisation: Polyphasic inspirations and expirations prevail during walk, trot and slow gallop, while mainly monophasic respirations were found when the horses were standing or galloping. Some horses were able to synchronize respiration and stride for short periods during pace and trot. During gallop all horses synchronized respiration to either a 1:1 or 1:2 ratio. A frequent shift between these two ratios was observed at slow gallop.

2. Steady state values: Minute volume increased lineari-ly with running speed and varies little at a given velocity. During rest, walk and trot the horses show considerable variability in the way this volume was produced. Either large tidal volumes at low frequency or small tidal volumes at higher frequency were found. Between rest and gallop V_T only doubled, while frequency increased 4 to 6 times. Variability of tidal volume was highest during rest and pace and was lowest during fast gallop in absolute as well as in relative figures. The degree of variability was in general similar in inspiration and expiration, indicating that both were active and equally well controlled.

3. Respiratory transients: The method is specially suited to show breath by breath the dynamic changes of depth, flow rate and duration at the beginning and the end of exercise. After trot or gallop respiratory frequency decreases immediately while tidal volume first increases, reaching values up to 24l, and then decreases.

References:

HÖRNICKE,H.; EHRLEIN,H.J.; TOLKMITT,G.; NAGEL,M.; EPPLE,E.; DECKER,P.; KIMMICH,H.P.; KREUZER,F.: Method for continuous oxygen consumption measurement in exercising horses by telemetry and electronic data processing; in:MENKE, LANTZSCH and REICHL: Proceedings of the 6th Symposium on energy metabolism of farm animals pp. 257-260 (Stuttgart 1974).

KIMMICH,H.P.; VOS,I.A. and KREUZER,F.: Telemetry of respiratory air flow; in: KIMMICH and VOS Biotelemetry (Meander N.V. Leyden 1972).

Supported by grant No. 11 1201 of Stiftung Volkswagenwerk Hannover. We are greatly indebted to Land gestüt Marbach for the use of their animals in these experiments.

Biotelemetry II. 2nd Int. Symp., Davos 1974, pp. 149–151 (Karger, Basel 1974)

Results of Radiotelemetric Measurements of the Energetic Output and Static Stress Using One-Man Saws

Věra Hanusová

Institute of Hygiene and Epidemiology, Praha

Development of radiotelemetric transmission of physiological parameters is of extraordinary importance for the physiology of work. In 1973 (1) we published a report on a miniature respirometer acting as a sensing unit of the respiratory volume for telemetric transmission. The respirometer, weighing 68g, was fastened on a half-mask before the respiratory valve. The volume of the passing air is thus transformed into electric impulses. The magnitude of an impulse equals to 200 mℓ of air up to ventilation of 100 litres/min. In this paper a report is made on the use of the respirometer in felling and trimming trees, our aim being to study reactions of the organism to statico-dynamic stress in different working positions. The first transmitter channel of the double-channel telemetric set, Teltest (Chirana Works), served to transfer impulses from the respirometer, and the second one, the R-waves of cardiac biopotentials. Ventilation, however, could be registered on flat ground up to the distance of 800 m. Moreover, samples of expired air were intermittently taken into two bags with a capacity of 2 ℓ. Taking samples took 1-2 min. Analysis of gases was effected by means of the interferometer. In total, 8 activities were studied and 133 measurements from an experienced and healthy worker were made. The measurements were effected at the temperature of 1-9° C. The static stress was evaluated by means of the index: WHR/Wkcal (2).

Fig. 1 shows the reflection of the working intensity and organism reaction in the HR and ventilation. In felling lasting 90 sec., the HR value increases from 107 to 130/min., ventilation from 24 to 32 STPD ℓ/min. In trimming, both the curves are lower than in felling; however, in the 9th minute after strenuously rolling-over logs lasting 24 sec., only the HR reached 140/min. and ventilation 30 ℓ/min. in the course of further trimming. Course of curves is an example of the variability of work under natural working conditions. In both functions the influence of previous activities distinctly persists both in the sense of increase and decrease. This phenomenon, detectable by means of continuous measurements and evaluations only, explains the considerable dispersion of values measured at the same activities. The curves show also that operations lasting a longer time enable the worker to retain a certain homeostasis in the output of calories as manifested by only a slight difference in mean values of different activities. We measured six principal and two minor activities: with strong trees of 0.5 to 1.0 m^3 and thin ones up to 0.1 m^3 of matter (Table I).

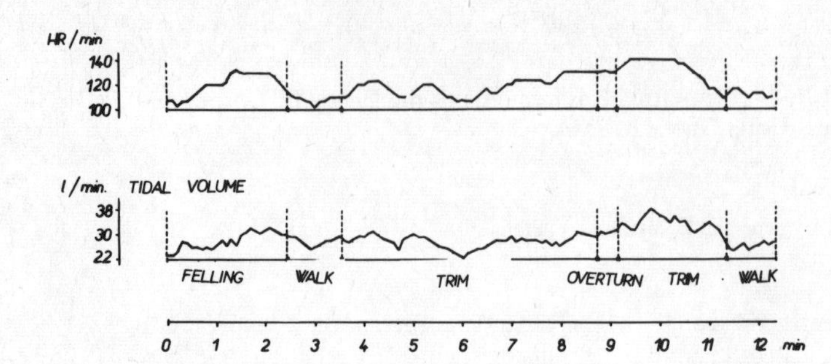

Fig. 1. Typical course of the HR and ventilation in felling and trimming trees with 0.8 m^3 of matter. The curves were plotted on the basis of moving averages from three intervals lasting 6 sec., converted to minute values.

Table I. Heart rate, energetic output and index WHR/W kcal in work with one-man saws.

	Total HR/min.		Working kcal/min.		Index static stress	
	$\bar{x}$	var. coef.	$\bar{x}$	var. coef.	$\bar{x}$	var. coef.
I.a	132.4	7.0	6.85	23.1	10.1	15.5
b	142.5	6.7	7.28	13.7	10.7	11.7
c	123.8	4.6	5.97	16.2	10.0	15.7
II.a	120.3	5.5	5.20	14.6	10.7	15.2
b	126.3	6.4	6.75	17.5	8.9	25.4
c	120.3	4.3	7.03	9.5	8.0	15.1
III.	136.5	5.6	6.65	4.1	10.8	8.3
IV.	139.4	5.9	6.93	21.4	11.3	23.3

Expl.: I. felling, II. trimming a) strong trees on flat ground b) strong trees on slant ground, c) weak trees on slant ground; III. rolling-over, IV. pulling-down suspended trees.

The minute HR in mean values was the highest in felling strong trees on slant grounds, i.e. 142/min. and in pulling-down suspended trees, i.e. 139/min., then in rolling over and felling trees on flat ground, i.e. 136 and 132 resp.. All these values significantly differed from trimming, i.e. 120 to 126/min. A significant difference was also found between felling strong and thin trees on slant ground. The index

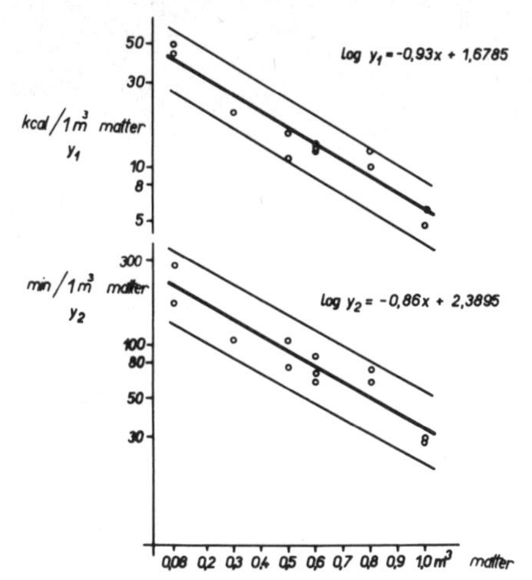

Fig. 2. Relation between mass of the manipulated trunks and a) time and b) energetic output on manipulation of 1 solid m^3 of matter.

of statical stress was the highest in pulling-down suspended trees, somewhat lower in rolling-over logs and felling on slant and flat ground. These values differed significantly from trimming on slant ground.

Working kcal/min. in mean values were approximately the same in all activities (6-7.3 kcal), the trimming value on flat ground being significantly lower (5.2 kcal). The differences mentioned above are caused by the working position and strength spent on the given activity.

While the minute energetic output for cutting strong and thin trees was approximately the same, there is a big difference in the energetic output per performance unit, i.e. 1 solid m^3 of matter. This is shown in Fig. 2 indicating the quantity of time and energy necessary for felling and trimming 1 solid m^3 of matter in relation to different mass of trees in the extent of 0.08 to 1.0 m^3 of matter. The dependance is exponential and therefore we used logarithmic expression for time and energy. The determined values were placed alongside calculated curves in the range within $\pm$ 2 sigma. Should this relation be confirmed later on, it will be possible to make use of it as a basis for physiologically substantiated standards.

Reference List:

1) HANUSOVÁ, V.: Ein Respirometer für kontinuierliche Messung des Atemvolumens mit Fernübertragung. Int.Z.angew.Physiol. 31, 141-150 (1973)

2) KLOTZBÜCHER, E.: Der Puls-/Energieumsatzquotient als Faktor zur Bewertung der Muskelermüdung unter Betriebsbedingungen. Zbl. Arbeitsmed. 8, 237-242 (1973)

Biotelemetry II. 2nd Int. Symp., Davos 1974, pp. 152–154 (Karger, Basel 1974)

Evaluation of Cardiovascular Drugs by Biotelemetry*

Ch. Scholand, R. Zerzawy and K. Bachmann
Medizinische Poliklinik, University of Erlangen, Erlangen

Pharmacological tests very often include results, investigated under controlled laboratory or clinical conditions. These results do not consider the possibly different effects of daily life activities. To prove whether there is a different effect or not we used a multichannel radiotelemetry system in 103 persons; the following three cardiovascular agents were investigated: Nitroglycerin, beta-blocking agents and Sympathomimetics.

Method: In every experiment continuous radiotelemetric measurement of direct arterial blood pressure(1), mean arterial pressure and heart rate was done.

Results:

1. Nitroglycerin: Direct arterial blood pressure and heart rate was measured by radiotelemetry in 18 patients with coronary heart disease. Systolic blood pressure and mean arterial pressure decreased at rest as well as during excercise statistically significant (Fig. 1). Diastolic blood pressure was slightly reduced. The increase of the heart rate, found at rest as well as during excercise, could be confirmed to be statistically significant only at rest before excercise. During excercise there was a remarkably small increase of heart rate.

2. Beta-blocking agents: Two beta-blocking agents were investigated, Propanolol and Pindolol, both applicated intravenously. Pindolol experiments were carried out in 34 patients with coronary

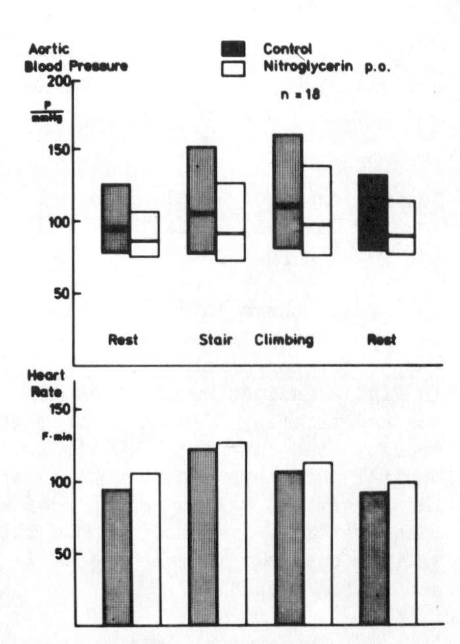

Fig. 1-Radiotelemetric measurement of arterial blood pressure and heart rate at rest and during excercise before and after Nitroglycerin p.o.

*Supported by a grant of the Deutsche Forschungsgemeinschaft

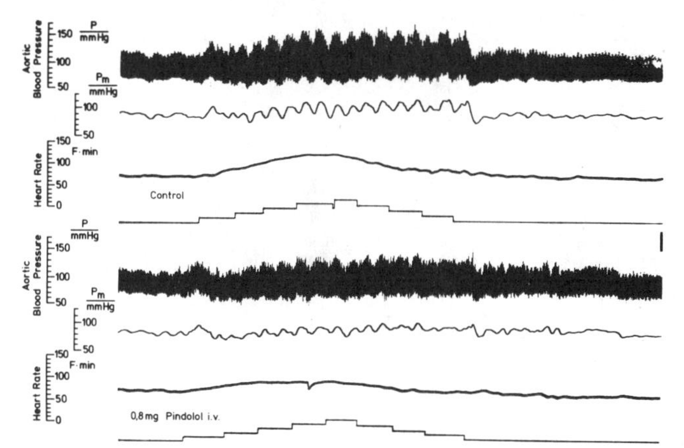

Fig.2-Continuous radiotelemetric measurement of arterial blood pressure and heart rate before and after Pindolol (0,8 mg i.v.)

heart disease.An original registration(Fig.2) demonstrates the main reactions of the cardiovascular system.The excercise hypertension is diminished and the excercise tachycardia markedly reduced.Propanolol,investigated in 38 patients with coronary heart disease,showed the same effects.Systolic and diastolic blood pressure and mean arterial pressure were reduced statistically significant.The decrease of the heart rate was found to be also statistically significant.The degree of decreased systolic blood pressure and heart rate under excise depends on the initial value (Fig.3).Dividing the entire group into two groups - a first one with an systolic blood pressure less than 160 mmHg,the second with a systolic blood pressure over 160 mmHg - Propanolol decreased the systolic blood pressure in the first group from 141 mmHg to 122 mmHg,in the second group from 177 mmHg to 147 mmHg.This difference was statistically highly significant.The same effect could be shown for the heart rate.At a heart rate over 120 beats/min the reduction is statistically significantly higher than at heart rate below 120 beats/min(from 133-111/min,from 105 - 96/min).

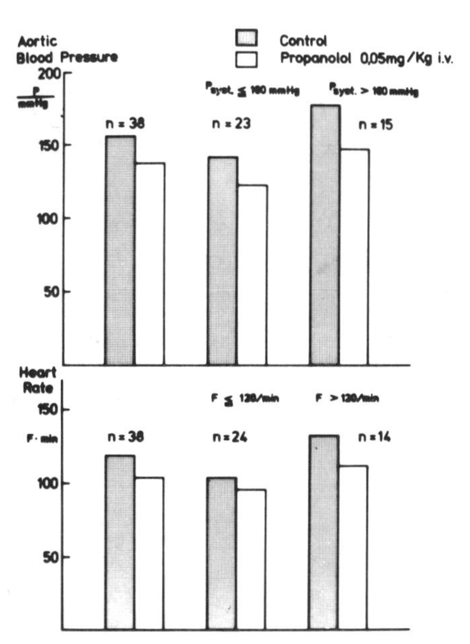

Fig.3 - Radiotelemetric measured systolic blood pressure and heart rate during excercise before and after Propanolol

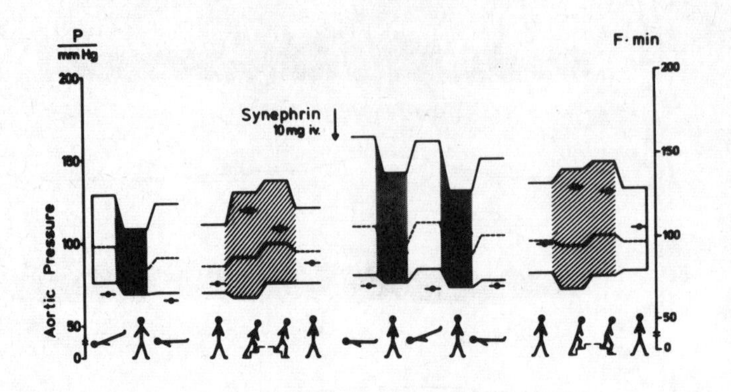

Fig.4 - Radiotelemetric measured arterial blood pressure and heart rate before and after Synephrine (10 mg i.v.)

3. Sympathomimetics: Synephrin, 10 mg i.v., was investigated in 13 patients with an orthostatic hypotension (Fig.4). The entire group showed in upright position a decrease of systolic blood pressure and mean arterial pressure, a moderate reduction of diastolic blood pressure and an increase of the heart rate. After application of Synephrine orthostatic dysregulation is more apparent; this difference is statistically significant for the systolic blood pressure and mean arterial pressure. During excercise after Synephrine appeared an augmented tachycardia in comparison with the control group.

Conclusion: Three different, widely used cardiovascular drugs were investigated in 103 patients by radiotelemetry. With this method we are able to observe the influences of daily life activities on the cardiovascular system. The results are in good agreement with those found experimentally. Therefore we believe, that radiotelemetric measurements of cardiovascular parameters are a helpful tool, getting information about the response of the cardiovascular system to physiologic, daily life activities.

References:

1 BACHMANN, K. and THEBIS, J.: Die drahtlose Übertragung kontinuierlicher direkter Blutdruckmessungen
Z. Kreislaufforsch. 56, 188 - 191, 1967

Biotelemetry II. 2nd Int. Symp., Davos 1974, pp. 155–157 (Karger, Basel 1974)

Intra-Corporeal Detector for Transplant Rejection

C. Fourcade, D. Cathignol, B. Lavandier, G. Dureau and J. Descotes
INSERM, Chirurgie Vasculaire et Transplantation d'Organes, Bron

We have shown that the electrical conductivity of a tissue reflects its state, particularly its viability and the intracellular and extra-cellular volume ratio.
During the rejection of a transplanted organ, there is commonly an extracellular oedema which induces a decrease in electrical resistivity of the parenchyma. This decrease seems to be earlier than the ECG modifications. By measuring the conductivity of a transplanted organ, it is thus theoretically possible to perform an early detection of the rejection, and thus to treat it successfully. On the other hand, this measurement could be a complementary test for liver and kidney which nowadays are monitored through biological tests only.
We offer an implantable device which measures tissue conductivity and transmits its value outside the body. A receiver analyses the signal obtained to display these results.

THE IMPLANTED TRANSMITTER : the schematic diagram of the transmitter and a commercial model coupled with a pace-maker are shown on figure 1.

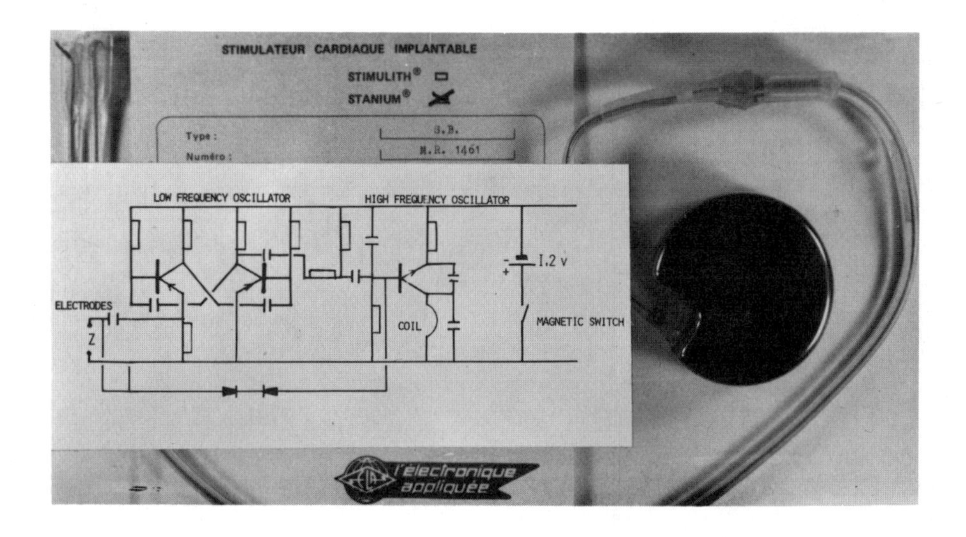

We are really measuring the modulus of impedance which is associated to the tissular electrical conductivity. Impedance is measured through an impedance-frequency converter obtained by a univibrator with a frequency of 8KHz. When the impedance to measure is connected in parallel on the emitter resistance, the oscillator frequency is shifted; by example 3KHz corresponds to 500Ω and 3.5KHz to 1500Ω. The low-frequency signal is then used for frequency modulation of a high frequency oscillator in the 88-108MHz band, where there is no local radio transmitter. As shown in figure 1, the high frequency signal is transmitted through the antenna constituted by the electrode wires themselves. The impedance of the tissue is measured through two standard cardiac pacing electrodes.
In order to save the batteries, a magnetic switch activates the device only during the measurement.

THE RECEIVER : The receiver's schematic diagram is shown on fig. 2. In order to receive only the signals coming from the impedance detector, a receiving loop is placed opposite to the transmitter. A permanent magnet placed on the loop antenna closes the magnetic switch. The antenna is connected to a standard frequency modulation receiver until the detection stage. Then the signal is treated and the pulses are converted into a tension which drives a galvanometer to display the modulus of intramyocardic impedance. When the tension is lower than a threshold limit value, an alarm is triggered.

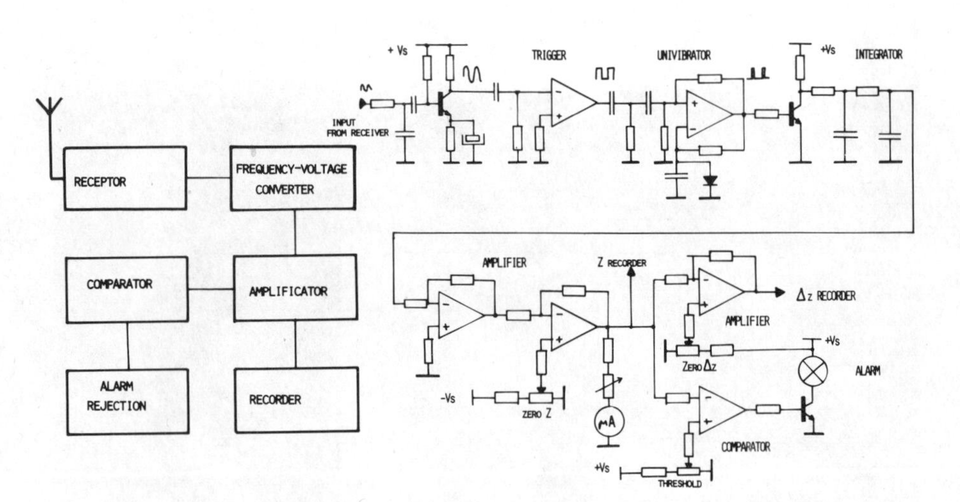

Figure 2 schematic diagram and alarm rejection system

RESULTS AND DISCUSSION : Fig. 3 shows:

-the evolution of intramyocardic impedance on a control dog; it is largely stable.
-the evolution on impedance of a non-treated animal only treated between the 7th and 9th days.
-the evolution for a treated animal : impedance decreases as soon as the treatment is discontinued.
The evolution on transplanted animals when the amplitude of the R wave on the ECG and the impedance are compared, shows the impedance decreases, reflecting a rejection reaction, before any change in the ECG.

Conclusion : By measuring myocardical tissue conductivity, early detection of rejection is possible. An implantable conductivity transmitter gives information upon the evolution of a transplanted tissue, allowing a rapid treatment.
This technique is useful in heart transplants, and can be used still more fruitfully in other organs, such as liver or kidney, where physical detection techniques such as ECG cannot be used.

Brevets n° 72 03 815
n° 73 06 129

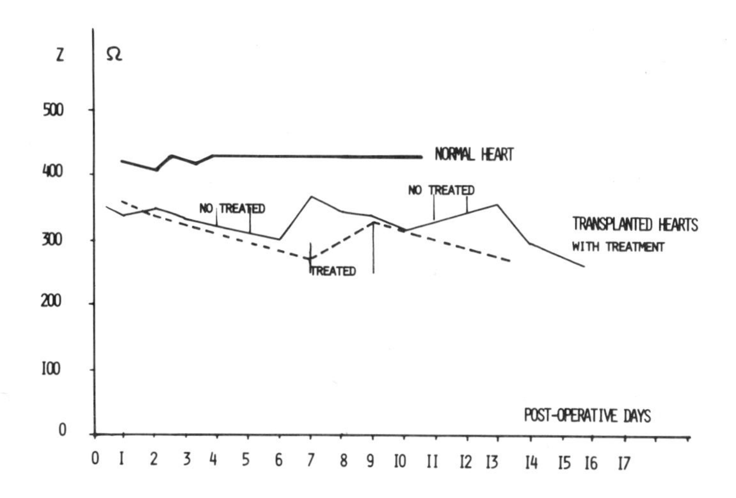

Figure 3 Intramyocardiac impedance for normal and transplanted dog

Biotelemetry II. 2nd Int. Symp., Davos 1974, pp. 158–160 (Karger, Basel 1974)

Instrumentation for Renal Hemodynamic Studies in Unrestrained Dogs

Roland D. Rader, Christopher M. Stevens, John P. Meehan and James P. Henry
University of Southern California, School of Medicine, Department of Physiology, Los Angeles, Calif.

The use of a chronically implanted telemetry system that measures renal artery blood flow and abdominal aortic blood pressure in totally unrestrained animal subjects has made it possible to begin assessing the role that renal hemodynamics plays in the development of naturally occurring or experimentally induced hypertension. A standard FM-FM system transmitting at a nominal frequency of 240 MHz and using IRIG subcarrier channels 7 and 9 is employed in combination with a command receiver that is activated by a 2.8 MHz RF signal, but which automatically turns off at the end of a preset interval. The package (Fig.1) including the battery weighs 100 grams and will operate continuously for approximately 100 hours. With the ability to control transmission time, this system can be used to obtain several hundred 10-minute recordings.

The carrier consists of a single stage LC oscillator occupying a volume of 1 cubic centimeter. A power dissipation of 10 mW permits reliable transmission to a distance of 60 meters. A metal shield that encloses the LC circuit acts as an antenna and also makes the oscillator relatively immune to frequency changes caused by placing the oscillator near larger objects. Each subcarrier consists of two transistorized parallel-T oscillators with individual frequencies of oscillation operating with common, collector and emitter resistors. When both transistors are conducting equally, the resultant frequency approximates the average of the two individual frequencies. Frequency modulation, which is limited between the two individual frequencies, is accomplished by controlling the base bias on either one or both of the transistors.

The command receiver contains four 2N5089 transistors, two of which are connected as an AC-coupled amplifier and used to detect and amplify the 800 Hz amplitude-modulated 2.8 Hz command signal. Of the remaining two transistors, one is used as an emitter-follower, and the other, as a DC amplifier. The DC amplifier connects to a silicon controlled switch that has a 5 to 10 minute on time.

Pressure is sensed by a miniature transducer (Konigsberg Instruments, Inc.) which is implanted in the abdominal aorta. By using a 1% duty cycle excitation, the power consumption in the bridge is kept at approximately 0.5 mW. After amplification, the bridge signal is filtered and used to frequency modulate a subcarrier oscillator which

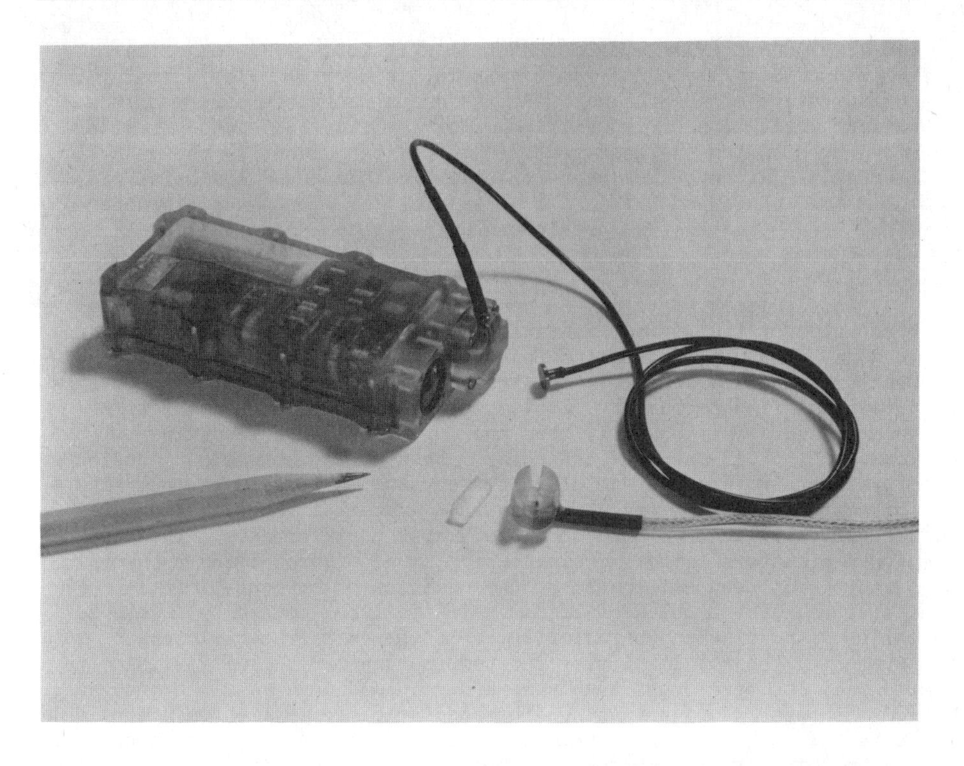

Fig. 1. Two-Channel Implantable Telemetry Package

in turn is used to synchronize the transducer excitation. A total power of 2.5 mW is dissipated in the blood pressure channel.

Blood flow in the left renal artery is detected by measuring the differences in phase between bursts of ultrasonic energy transmitted upstream and downstream within the blood vessel (1). Line illumination across the diameter of the vessel yields a phase difference between upstream and downstream energy proportional to the volume flow rate, i.e., the product of vessel cross-sectional area and blood flow velocity, whereas, point illumination through the center of the vessel yields a phase difference proportional to the volume flow rate divided by the radius. Variation of the diameter in point illumination produces an alteration in the sensitivity, but little change in the zero flow baseline, whereas, with line illumination, a change in the diameter causes only a change in zero flow baseline. The particular significance of line illumination is that the changes in diameter caused by changes in vascular tone and pressure fluctuation are incorporated directly into the sensitivity of the flow measurement.

A renal model that considers the regional pressure drops and storage capacities within the renal vasculature is used in assessing renal responses (2). Preglomerular flow resistance, a composite resistance of all preglomerular vessels, is determined by dividing the maximum value of the derivative of abdominal aortic pressure by the maximum derivative of kidney flow. Total kidney resistance is determined by dividing the mean pressure across the kidney by the mean flow through the kidney. The postglomerular resistance, a composite of resistance contained within all postglomerular flow paths, is determined by subtracting the preglomerular resistance from the total resistance.

This telemetry system, in conjunction with the renal model, has been used for the past two years to obtain and analyze renal responses in dogs. The majority of the failures have been associated with the pressure and flow sensors. In these studies, both the pressure transducers and flow probes have lasted up to 12 months. Pressure transducer failure has generally been due to a break in a lead wire or leakage of fluid into the connector. Lead wire breakage and erosion of the lead wire solder joint at the crystals have been responsible for most of the flow probe failures. A number of improvements have recently been incorporated to improve system performance. The occasional premature battery drain, caused by a failure of the turn-on circuit, has been eliminated by the addition of current-limiting resistors. The problem of transmission failure, caused by a deterioration of modulation characteristics of the carrier oscillator, apparently the result of elevated temperature and humidity within the package, has been considerably retarded by adding an auxiliary package for the battery and using the liberated space for a lithium hydroxide desiccant. The use of stainless steel lead wire and isolation of the solder joint at the crystal from body fluids by a thin plastic sheet appear to have extended the lifetime of the flow probe.

These improvements should permit this telemetry system to acquire renovascular data from unrestrained dogs for longer than our past record of 5 months.

References

RADER, R. D.: Cardiovascular telemetry implants. Telemetry J. 6 15-20 (1971).

STEVENS, C. M. and RADER, R. D.: Telemetered renal blood pressure and flow responses to situational stress in unrestrained dogs; in IBERALL and GUYTON, Regulation and control in physiological systems; pp. 427-429 (Instrument Society of America, Pittsburgh, 1973).

This work was supported by the United States Air Force, Air Force Office of Scientific Research, Grant No. AFOSR-72-2190.

Biotelemetry II. 2nd Int. Symp., Davos 1974, pp. 161–165 (Karger, Basel 1974)

Use of Ultrasound Telemetry to Monitor Fetal Heart Activity

Orvan W. Hess and Wasil Litvenko
Yale University School of Medicine, New Haven, Conn.

The introduction of electronic techniques for detection and elucidation of signals derived from variations in electrical energy generated in the course of body processes has been one of the notable advances in the past decade. Among the vast variety of applications of such apparatus in clinical medicine, continuous electronic recording of the fetal heart activity is an outstanding example.

The use of ultrasound for the detection of fetal heart activity (MOSLER 1972) has broadened the range of applications of electronic techniques employed for monitoring the status of the fetus during pregnancy and labor. With appropriate sensors placed on the abdominal wall, it is now possible to obtain a continuous record of the cardiac rate of the fetus prior to rupture of the amniotic sac when direct attachment of the electrodes to the fetus is impossible or, for other reasons, may not be feasible. Reliable information which may be procured in this manner and measured against universally accepted indices of well-being is essential for the assessment of the condition of the fetus. Concomitant monitoring of the fetal heart rate and uterine contractions adds significantly to the safety of induction or augmentation of labor. Surveillance of these parameters in much the same manner as in an intensive care unit has been shown to have special value in the clinical management of patients when high risk of compromise of fetal reserve exists. It also provides means to observe the fetal cardiac response to medications, anesthetic agents or other factors which may have a deleterious effect. Alteration in patterns of instantaneous heart rate permit early detection of signs of fetal distress (HON 1968) and treatment of underlying causative factors.

In previous communications, we have described the development of miniaturized apparatus and a system for radio-telemetering (HESS 1962) the fetal electrocardiogram when obtained by standard methods with electrodes placed externally on the abdominal wall or applied directly to the fetus. The purpose of the present communication is to report the development of electronic apparatus for telemetering of continuous fetal heart activity (Fig. 2) detected by ultrasound. A system which has been developed for simultaneous continuous telemetry of uterine activity (Fig. 3) during labor is also described. A series of selected tracings which demonstrate the type of recordings obtained in a wide variety of circumstances is presented.

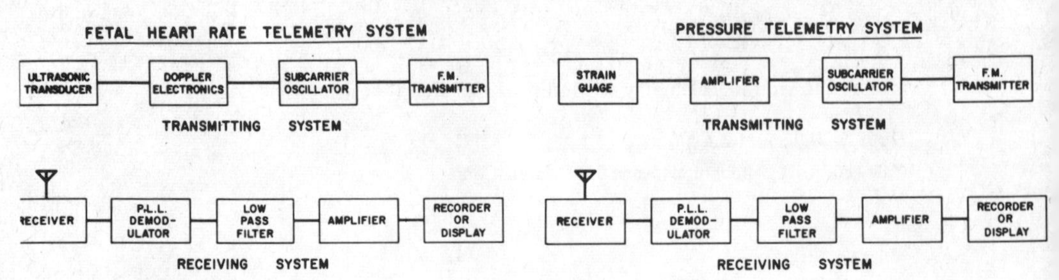

Fig.2 Diagram of system for radio-telemetry of heart activity detected with ultrasound.

Fig.3 Diagram of system for radio-telemetry of uterine activity detected with tocograph.

Methods and Materials. To delineate fetal heart activity, the type of ultrasound transducer (Corometrics, Inc., Wallingford, Conn.) which is usually employed consists of a light weight circular disc containing two or more transmitting and receiving crystals. From these crystals, the ultrasound vibrations are transmitted through the abdominal wall at an angle of approximately 4-5 degrees at a frequency of 2 MHz with an intensity of about 7-10 milliwatts. The total radiated energy delivered to the fetus is estimated to be less than 2 milliwatts/ sq. cm. which assures safety to the mother and to the fetus. Several investigators (BERNSTINE 1969) have reported that no microscopic injuries were observed with even higher levels of exposure to ultrasound.

Uterine activity associated with contractions of labor is detected by means of the usual tocographic sensor (Corometrics, Inc.) with modification to allow flexibility in range of sensitivity. This is placed externally on the abdominal wall usually at the level of the uterine fundus. The pressure sensitive transducer permits recording of the onset and duration of uterine contractions. The resultant curve of the contraction does not represent the actual intrauterine pressure. However, in many instances there is a close correlation with that noted by internal catheter methods.

The block diagram (Fig. 1) displays the overall layout of the important or significant components of the transmitting and receiving equipment. The telemetry system has been designed to operate in the 146.94 MHz band employing frequency modulation transmission. Channelized crystal frequency control is used in both the receiver and transmitter operating at a 15 kHz deviation.

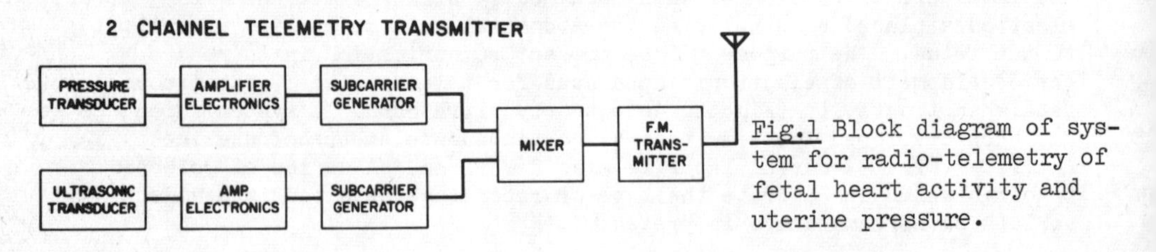

Fig.1 Block diagram of system for radio-telemetry of fetal heart activity and uterine pressure.

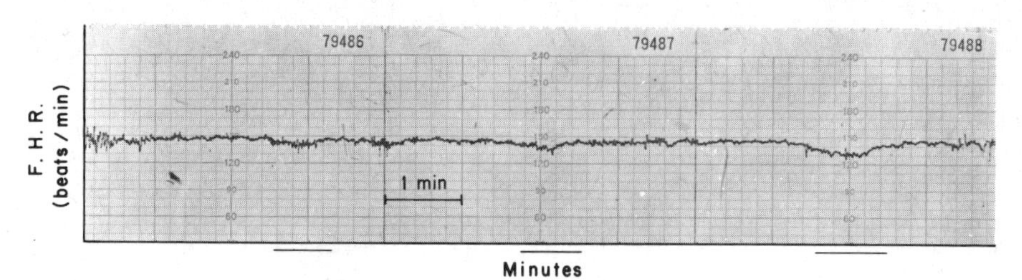

Fig.4 Tracing of fetal heart rate telemetered during use of ultrasound. Patient in late labor exhibiting mild bradycardia with pattern suggestive of head compression.

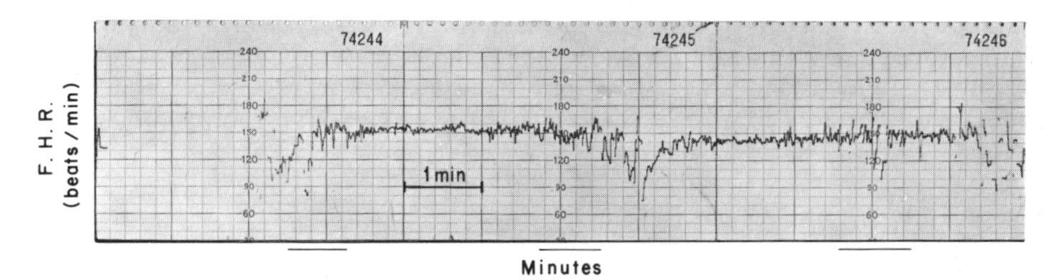

Fig.5 Tracing of fetal heart rate telemetered during use of ultrasound. Patient in active labor exhibiting bradycardia with uterine contractions suggestive of umbilical cord compression.

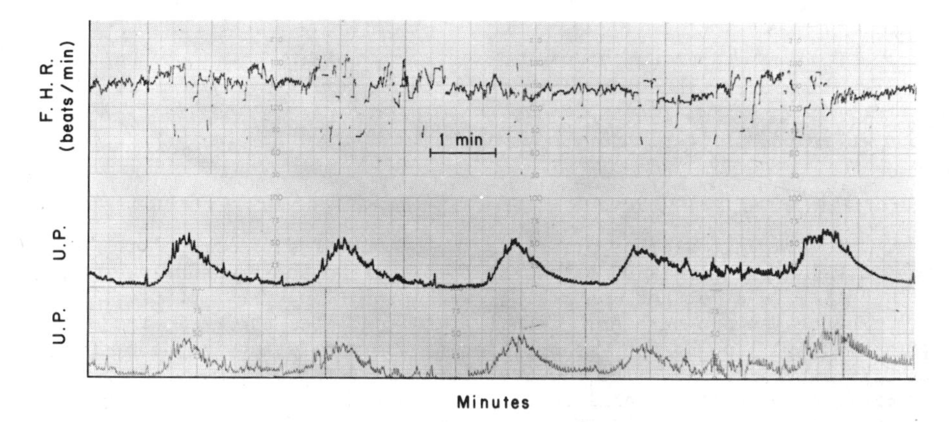

Fig.6 Simultaneously recorded tracings of fetal heart rate and uterine activity during active labor. The upper tracing is the f.h.r. The middle tracing of uterine contraction patterns was recorded in the usual external technique for comparision with that recorded at the same time using telemetry.

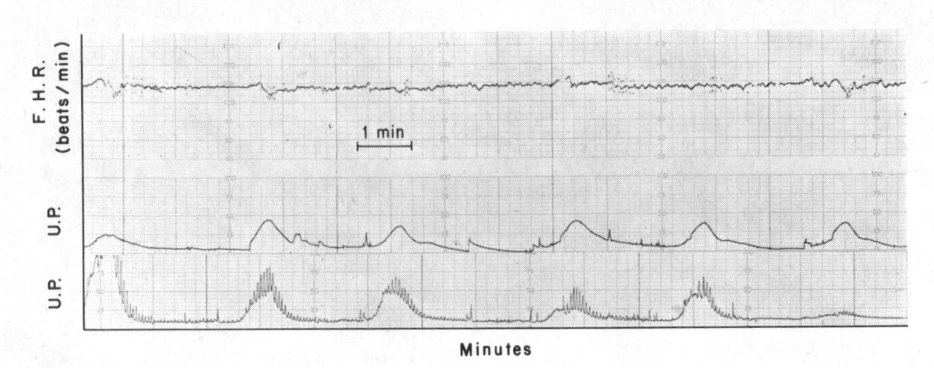

Fig.7 Simultaneously recorded tracings of fetal heart rate and uterine contractions. The upper tracing is the f.h.r. The middle tracing of uterine contractions was recorded with use of internal (catheter) technique. The lowermost tracing was recorded during telemetering. The progressive decrease noted in amplitude of the contraction curves represents use of the built-in adjustable sensitivity control.

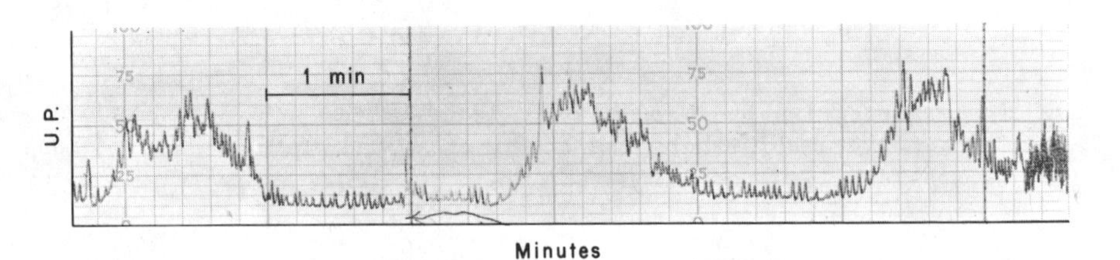

Fig.8 Tracing of pattern of uterine activity obtained from abdominal wall using tocographic technique after telemetering approximately 300 feet. The sharp intervening peaks noted on the uterine contraction curves represent respiratory movements.

To transmit uterine pressure, the output of a tocograph strain gauge which produces a DC output proportional to the pressure exerted is applied to a voltage controlled oscillator (VCO) for amplification to produce the sub-carrier. This in turn modulates the radio-frequency transmitter. 160 Hz/sec was selected as the modulating frequency for the pressure channel. The receiver portion of the system is a conventional, commercially available, highly sensitive (.5mv for 20db dual conversion) Kenwood KP202 FM receiver which has been modified as indicated to permit it to discriminate the sub-carrier signal. A phase-lock loop (PLL) type demodulator accomplishes this function. The signal may be further filtered and amplified as required to permit it to operate a recorder (Clinical Monitor-101B, Corometrics, Inc.) or some type of display unit.

The fetal heart activity is detected on the basis of the Doppler principle by means of the ultrasound instrument via the incorporated transmit and receive crystals and the additional electronics associated with the unit. As in the uterine pressure system, the electrical signals obtained in this manner are fed into a sub-carrier oscillator, modulated and in turn serve to drive the FM transmitter.

The two sub-carrier oscillator outputs are combined and multiplexed to permit the two-channel signal to be transmitted. Possibilities exist for adding other sub-carriers if desired. The RF portion of the transmitter consists of a crystal oscillator which is varactor modulated, and operates at 48.98 MHz. This is followed by a tripling stage which places the transmitter on 146.98 MHz. The output is amplified further to produce about one-half watt of power. This was found to be useful for penetrating difficult areas where thick walls or partitions exist. Transmission has been successfully accomplished at distances of 300-500 feet between transmitter and receiver stations.

Conclusions. Radio-telemetry offers a useful alternative method for clinical monitoring of fetal heart rate and uterine contractions.

A system and instrumentation have been developed for continuous telemetry of fetal heart activity detected by ultrasound.

A system has been devised for multiplexing and simultaneous telemetering of signals associated with changes in uterine pressure.

References

Bernstine, R. L.: Safety studies with ultrasonic Doppler technique. Obs.-Gyn. 34: 707 (1962).

Hess, O. W.: Radio-telemetry of fetal heart energy. Obs.-Gyn. 4 :516-521 (1962).

Hon, E. H.: An atlas of fetal heart rate patterns. (Harty Press, Inc., New Haven, 1968).

Mosler, K. H.: A new ultrasonic monitor for use in obstetrics. Internat'l J. of Obs.-Gyn. Vol. 10, No. 1: 1 (Jan. 1972).

Session VII

Telemetry of Neurobiological Parameters

Chairmen: *A.A. Borbély, G. Dumermuth and J.A.J. Klijn*

Biotelemetry II. 2nd Int. Symp., Davos 1974, pp. 168–172 (Karger, Basel 1974)

Monitoring Neurobiological Processes: Cable Connections or Telemetry?*

Alexander A. Borbély
Institute of Pharmacology, University of Zürich, Zürich

Introduction Traditionally, cable connections are used to record from electrodes which have been implanted in the brain or in other structures of freely moving animals. The cable may be connected to swivel joints to permit a maximum amount of free movement, and to prevent the leads from twisting. Similarly, scalp electrodes in man are usually connected via cables to the EEG-amplifier. The recent development and increasing availability of telemetric methods has made it possible to monitor neurobiological processes in entirely unrestrained animals and men. A recent survey shows that in 1973 52 firms have been marketing telemetry systems in the United States (Guide to Scientific Instruments 1973-1974. Science 182A: 138-139 (1973)). Moreover, on the basis of published circuit diagrams, even the non-expert can successfully attempt to build transmitters for biological signals. Therefore, it may be tempting to foresake the classical methods of monitoring the brain and behavior in favour of telemetric techniques. The ensuing short discussion is based on my own experience with telemetry, and will focus on some problems which may be relevant for the applications of the method.

Experiments with small animals Small laboratory animals such as rats or mice are widely used in neuropharmacological and behavioral research. They are easy to handle and to keep, are inexpensive, require little space, and show a wide repertoire of behaviors. Their brain chemistry has been intensively studied, and is in many respects comparable to that of higher mammals. In this laboratory telemetry has been used for a number of years to study pharmacological and environmental effects on the regulation of body temperature in mice (GRAF and BORBÉLY, 1966) and rats (BORBÉLY, 1970; BORBÉLY, BAUMANN and WASER, 1972; BORBÉLY and HUSTON, 1972, 1973; BORBÉLY, BAUMANN and WASER, 1974). Implantable, battery-powered transmitters

* Supported by SNSF, grant nr. 3.212.73.

are used to record continuously intraperitoneal temperature during long-term experiments. After implantation, the peritoneal cavity is closed completely, thus minimizing the risk of infection.

The ECG may serve as a further indicator of influences of emotional stimuli or drugs on the brain, and can be recorded in the rat with relative ease by telemetry (e.g. BORBÉLY,BAUMANN and WASER,1972; BOHUS, this volume).

The EEG may serve to characterize the functional state of the brain and to detect changes in specific brain areas. Due to the small dimensions of the rat's head, few attempts have been made so far to record brain potentials by telemetry in this species. Recently, a miniaturized 4-channel EEG-transmitter has been developed (VOEGELI and KRAFT,1972; KRAFT and VOEGELI,1973) and successfully applied in physiological and pharmacological experiments (BORBÉLY, BAUMANN and WASER,1972; BORBÉLY and HUSTON,1973; BORBÉLY, HUSTON and WASER,1973; BORBÉLY et al.,1973; BORBÉLY and HUSTON, in press). A miniaturized accelerometer can also be incorporated in the transmitter system to monitor the rat's head movement pattern (DÄNIKER,1973; BORBÉLY et al., 1973).

In most of the studies mentioned above, auditory evoked potentials were recorded along with the spontaneous EEG. However, the use of sensory stimuli for eliciting evoked potentials is not without problems, since freely moving animals vary their orientation to the stimulus source. This difficulty can be avoided by using electrical stimulation of peripheral or central nervous structures instead of sensory stimuli. Evoked potentials elicited by telestimulation and recorded by telemetry have been studied in behaving cats (SPERRY et al.,1968).

Two major problems in the use of multi-channel EEG-telemetry in rats should be mentioned: (1) The limited life-span of the transmitters, varying from several weeks to more than a year. Although the causes of such failures are being recognized and the problems gradually solved, the long-term reliability of any newly developed, miniaturized system should not be assessed with undue optimism. (2) A firm and reliable fixation of the transmitter assembly on the rat's relatively thin skull is difficult to achieve.

In summary, recent progress in miniaturization of electronic components and circuits has opened the possibility of using multi-channel biotelemetry even in small laboratory animals.

Do cable connections influence behavior ? The monitoring of neurobiological processes via cable connections is for many experimental purposes an adequate technique. However, in some instances, cable leads may influence the animal's behavior and thus intefere with the experiment. BOHUS (this volume)has clearly shown that the heart rate response of rats to emotional stimuli adapted more rapidly, when telemetric methods were used. Moreover, the resting heart rate of a series of mammals recorded by telemetry was approximately 25 % lower than the corresponding values obtained with classical methods (ESSLER,FOLK and ADAMS, 1961). Also the rat's sleep behavior differed significantly, if recordings via cable connections and via telemetry were compared (HAUPT,1973). Both, the animals' slight impediment of movement, and the intermittent visual (and possibly acoustic) stimulation by dangling cable leads, could account for these differences. In another interesting experiment using EEG-telemetry it was shown that the sleep of horses and cattle recorded under field conditions differed markedly from sleep under stall conditions (TOUTAIN and RUCKEBUSCH,1973; RUCKEBUSCH,DALLAIRE and TOUTAIN, in press). Thus the use of telemetry makes it possible to study the long-term behavior of animals in their natural environment or habitat. Similar studies in man would be particularly interesting. However, first it will be necessary to overcome the problem of obtaining reliable, artifact-free EEG-records with scalp electrodes from freely behaving subjects. It is also worth remembering that in man as well as in larger animals, miniaturized portable recording devices (see McKIMMON, this volume) may constitute a valuable alternative to telemetry.

In summary, the few controlled studies indicate that the use of cable connections may indeed influence sleep and emotional responses in animals.

Experiments necessitating telemetry   Studies which are not feasible without the use of telemetry, include experiments in animals in their natural environment (particularly aquatic or hibernating animals, birds) or under extreme conditions (e.g. extreme cold or heat, high *g* environment, radioactive environment, animals and men in space). The monitoring of the adaptation of the organism as well as the study of biological rhythms is often a main objective of such studies. Several intersting applications of telemetry in animal biometeorology have been summarized in a recent review article by FOLK and COPPING (1972).

A further area of research necessitating the use of telemetry and/or telestimulation is the study of group behavior and social interaction. Here the pioneering studies by DELGADO (1964,1970) and MAURUS and PLOOG (1971) of the social behavior of monkeys deserve mention. The possibility of monitoring neurobiological processes in a natural social setting will undoubtedly allow fruitful applications of telemetry in a variety of experimental situations. Thus, the era in which neurobiological research in animals and man was limited to the study of individuals kept in isolation under highly artificial laboratory conditions, may soon come to an end.

Concluding remarks For all those considering the use of telemetry for monitoring the brain and behavior it may be useful to remember that practically all the recent progress in neurobiology has been made without these sophisticated recording methods. Although the application of biotelemetry may considerably extend the scope of experiments, relevant data can still be obtained by recording via cable connections. The decision to use telemetric methods, whether out of necessity or for sheer elegance, involves often more than just adopting some new technique. It may be necessary to develop new methods for fastening the transmitter to the experimental animal so that neither he nor his curious mates can gain undue access to the new device; it may necessitate modifying the cages for placement of the antennas and shielding off sources of radio-interference; it may involve a substantial amount of trouble-shooting, and the recognition and elimination of unusual artifacts in the records. These laborious and time-consuming aspects of telemetry should be given serious consideration before committing oneself to this technique. It should be remembered that telemetry is truly problem-oriented: it may help solving new problems, and it will most certainly create new ones.

References

BORBÉLY,A.A.: Telemetry of body temperature, motor activity, food and fluid intake in the rat; in Proceedings of 4th International Congress on Pharmacology; Vol.5,pp.295-297 (Schwabe,Basel 1970).

BORBÉLY,A.A., BAUMANN,I. and WASER,N.M.: Multichannel telemetry of physiological parameters (body temperature, ECG,EEG) in the rat. II.Applications in neuropharmacology; in KIMMICH and VOS Biotelemetry; pp.381-388 (Meander,Leiden 1972).

BORBÉLY,A.A. and HUSTON,J.P.: Gamma-butyrolactone: an anesthetic with hyperthermic action in the rat. Experientia 28: 1455 (1972).

BORBÉLY,A.A.,DÄNIKER,M.,MOSER,R. and WASER,P.G.: Multiparameter telemetry in neuropharmacological research. AGEN-Mitteilungen 15: 29-34 (1973).

BORBÉLY,A.A. and HUSTON,J.P.: Effects of gamma-butyrolactone on body temperature and evoked potentials in the rat; in KOELLA and LEVIN Sleep: Physiology,Biochemistry,Psychology,Pharmacology,Clinical Implications;pp.355-359 (Karger,Basel 1973).

BORBÉLY,A.A.,HUSTON,J.P. and WASER,P.G.: Physiological and behavioral effects of parachlorophenylalanine in the rat. Psychopharmacologia 31: 131-142 (1973).

BORBÉLY,A.A.,BAUMANN,I.R. and WASER,P.G.: Amphetamine and thermoregulation: studies in the unrestrained and curarized rat. Arch.exp.Path.Pharmakol. 281: 327-340 (1974).
BORBÉLY,A.A. and HUSTON,J.P.: Selective enhancement of slow-wave sleep by light in the rat; in KOELLA and LEVIN Proceedings of Second European Congress on Sleep Research (Karger,Basel in press).
DÄNIKER,M.H.: Subminiature accelerometer for radiotelemetric recording of motor activity in the rat; in Proceedings of the 10th Conference on Medical and Biological Engineering; p.61 (Dresden,1973).
DELGADO,J.M.R.: Free behavior and brain stimulation. Int.Rev.Neurobiology 6: 349-449 (1964).
DELGADO,J.M.R.: Telecommunication in brain research; in Proceedings of 4th International Congress on Pharmacology; Vol.5, pp.270-278 (Schwabe,Basel 1970).
ESSLER,W.O.,FOLK,G.R. and ADAMSON,G.E.: 24-hr cardiac activity of unrestrained cats. Fed.Proc. 20: 129 (1961).
FOLK,G.E. and COPPING,J.R.: Telemetry in animal biometeorology; in TROMP,WEIHE and BOUMA Biometeorology; Vol.5, Part II, pp.153-170 (Swets and Zeitlinger, Amsterdam 1972).
GRAF,H. and BORBÉLY,A.: Radiotelemetrische Körpertemperaturmessung bei Mäusen und Ratten. Experientia 22: 339-340(1966).
HAUPT,R.: The sleep-wake behavior of rats. A comparison of cable and telemetric registration; in KOELLA and LEVIN Sleep: Physiology, Biochemistry, Psychology,Pharmacology, Clinical Implications; pp. 285-289 (Karger,Basel 1973).
KRAFT,W. and VOEGELI,F.: 4-Kanal Miniatursender zur Uebertragung des Elektroencephalogramms von Kleintieren. AGEN-Mitteilungen 15: 19-24 (1973).
MAURUS,M. and PLOOG,D.: Social signals in squirrel monkeys: analysis by cerebral radio stimulation. Exp.Brain Res. 12: 171-183 (1971).
RUCKEBUSCH,Y.,DALLAIRE,A. and TOUTAIN,P.L.: Sleep patterns and environmental stimuli; in KOELLA and LEVIN Proceedings of Second European Congress on Sleep Research (Karger, Basel in press).
SPERRY,C.J.,BACH,L.M.N.,HAPPEL,L.T. and CABANISS,J.S.: Implantable stimulator and transmitter for telemetry of evoked potentials during defensive behavior. Biomedical Sciences Instrumentation 4: 119-124 (1968).
TOUTAIN,P.L. and RUCKEBUSCH,Y.: Sommeil paradoxal et environnement. C.R.Soc.Biol. 167: 550 (1973).
VOEGELI,F. and KRAFT,W.: Multichannel telemetry of physiological parameters (body temperature,ECG,EEG) in the rat. I. Design and methods; in KIMMICH and VOS Biotelemetry; pp. 371-380 (Meander, Leiden 1972).

Biotelemetry II. 2nd Int. Symp., Davos 1974, pp. 173–175 (Karger, Basel 1974)

An Eight Channel Semi-Implantable Telemetry System for Animal Research

D.E. Olsen, S.L. Moise, jr. and S.W. Huston

Space Biology Laboratory, Brain Research Institute, University of California, Los Angeles, Calif.

Obtaining physiological data from unrestrained and uncooperative animals requires packaging technique that, irrespective of circuit design, is an art in itself. This is particularly true with monkeys who have highly manipulative hands that can pick and tear at even the most cleverly designed telemeters. At extremes of the data acquisition task are (1) restraining the animal in a collar attached to the cage and (2) a totally implanted telemetry system. The disadvantages of the former are obvious, however the requirement of surgery to exchange batteries in a total implant is hazardous and inconvenient as well. Moreover, for animal studies, it is impractical to propose a scheme in which power is transmitted through the skin to a passive implant.

The system described here (Figure 1) combines the best features of a hardwire system with the total mobility of an implant. Six EEG and one EOG channels are multiplexed along with animal ground, full scale and zero scale sync pulses. The biotelemeter is comprised of amplifiers, multiplexer and transmitter. The circuit design represents the improvement, over several years, of circuits previously described by OLSEN et al. (1971) and for that reason the circuit description is

Fig. 1. Animal With Telemeter Attached

brief. The amplifier design consists of three operational amplifiers (Siliconix L 144C1). The first two are input stages in the voltage follower configuration. Their outputs are capacitively coupled to the third, a differential gain stage. This instrumentation amplifier configuration, described by GODDEN (1969) gives input impedance of > 50 Mohms, common mode rejection ratio of > 90 dB and noise referred to the input of < 2 μv pk-pk. It should be noted that the operational amplifiers have to be screened for low popcorn noise. Current drain for each composite amplifier is set through an external resistor to approximately 20 μamps. The multiplexer consists of a multivibrator (clock) formed by two RCA CD 4001 inverters and associated passive components, a decade divider (CD 4017) and three analog gates (CD 4016). The integrated circuits (IC's) are complementary symmetry metal oxide semicomductors (CMOS) which account for the extremely low current drain of the multiplexer (250 μamps). It should be noted that the power alloted to the multiplexer is substantially higher than previously reported (7 μamps), however studies of the stability of the clock frequency have indicated that the increase is needed. The transmitter consists of a Vackar configuration oscillator followed by an output buffer amplifier which provides gain and isolates the antenna from the oscillator. FM modulation is accomplished by a varactor diode in the base circuit of the oscillator that is fed by the PAM wavetrain output. The transformers are wound on torroidal cores and are mounted directly beneath and on JFD MT 309 trimmer capacitor leads. Total system power drain is 2.4 ma from four series connected mercury cells (Mallory 675-180 MAH). Operating life of the system between battery changes is approximately 100 hours.

The amplifiers, multiplexer and transmitter are packaged as separate welded cordwood modules which, by using flat pack IC's, 1/8 watt resistors and miniature capacitors, are made quite small. The modules are mounted on a miniature printed circuit board along with four mercury cells which supply power. The total assembly is 6.8 cm by 3.6 cm by 1.7 cm, and weighs 54 grams, including batteries. The inverted assembly (printed circuit side up) (Figure 2) mounts in a lucite receptacle which is attached to the monkey's skull with dental acrylic. At one end of the receptacle is an 18 pin strip connector which mates with a corresponding connector at one edge of the printed circuit board.

The EEG electrode implantation procedure is standard. The EOG leads, silastic coated stainless steel wire, are routed subcutaneously to just above the suborbital rim. The electrode leads are then soldered to the appropriate connector leads, the telemeter is inserted into the receptacle and the system is attached to the skull with a bridge of dental acrylic. Three leads, shorted to each other on the animal side of the connector, are used to switch power to the system when it is installed in the receptacle. The antenna, a loop of wire surrounding the case, is then attached.

The transmitted signal may be received by any reasonable quality FM tuner. Of course, signal range may be enhanced with improvements

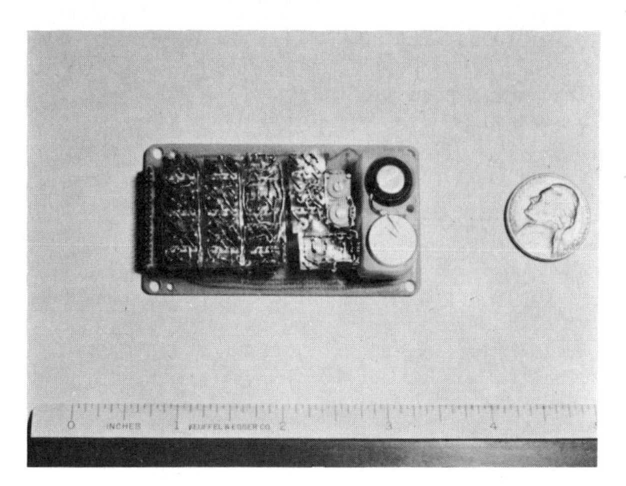

Fig. 2. Close-up of Electronic Package

in the receiving antenna and receiver. The received signal is then demodulated and reconverted to its original analog form using a model PAM-8 decommutator (BIOTEL).

The authors believe that the significance of this system is in its extremely small size, low power consumption and in the packaging methods used. Although the data acquired through use of the system is of excellent quality, several improvements are being made. First, is inclusion of either a magnetic or an RF switch for activating and deactivating the unit. Second, is optimization of the transmitting antenna and third, is further development of the surgical attachment technique of the receptacle to reduce size of the acrylic bridge and thereby lower the profile of the completed unit. In the future, a colony of unrestrained animals free to intermingle, each with his own implant, is envisioned and a hands off (excepting periodic veterinary examinations) behavioral study conducted.

GODDEN, A.K.: Amplify biological signals with IC's. Electronics Design. 17: 218-224 (1969).

OLSEN, D.E., FIRSTENBERG, A., HUSTON, S.W., DUTCHER, L.R., and ADEY, W.R.: An eight channel micropowered PAM/FM biomedical telemetry system; Record of the IEEE National Telemetering Conference. 308-310 (1971).

This investigation was supported in part by Public Health Service Research under Grant GM 16058-08 and in part by the AF Office of Scientific Research of the Office of Aerospace Research under Contract F44620-70-C-0017.

Biotelemetry II. 2nd Int. Symp., Davos 1974, pp. 176–178 (Karger, Basel 1974)

Experiences with a Telemetric System Permitting Simultaneous EEG Recordings and Brain Stimulation in Cats

P. Polc and H. Wolfgang

Department of Experimental Medicine, F. Hoffmann-La Roche & Co., Ltd., Basle

Telemetry is now a widely accepted technique in brain research, particularly useful in neuropsychopharmacological studies in animals (HIMWICH, KNAPP and STEINER (1965); BORBÉLY, BAUMANN and WASER (1972)). The aim of our investigation was to develop a system containing both the equipment for telemetering EEG and EMG data and a device for telestimulation of the brain in cats.

The telemetry system used in our experiments consists of a 4-channel subcarrier AM/FM transmitter and a stimulation receiver; its weight is 50 g. Since the whole equipment is mounted on the head of a cat, no extreme miniaturization, particularly of the power supply, was required in the development of the system. Thus, with a power consumption of ca. 800 μA and a NiCd battery of 60mA/h capacity, continuous recordings of more than 50 hours were provided without recharging the battery.

Fig. 1 shows the basic electronic circuit design of the system. Signals from stainless steel bipolar concentric electrodes implanted in the subcortical areas, as well as from cortical screws and an EMG electrode, are fed into either of 4 low noise differential amplifiers, which modulate the amplitude of 4 subcarrier generators by means of modulators. The generators oscillate at frequencies of 19; 16; 12.5 and 10 Kc, respectively. The multiplexed signal modulates the frequency of the transmitter in the range 100 - 108 Mc.

Some technical data of the transmitter and the receiver are given in Table I. In contrast to most other equipment , in our telemetry system the electrodes are connected to DC amplifiers of low input resistance (47 KΩ) without coupling capacitance. Potential drop and frequency dependence of the signal amplitude caused by this procedure are compensated in the system. In this way we obtain a high stability of the recordings against possible disturbances, due to sudden movements of the animal or changes in the electromagnetic field. For this purpose, however, a great precision in the construction of the electrodes is required. Particularly, an exact symmetry of the electric characteristics of both poles of an electrode is required in order to minimize tip potential.

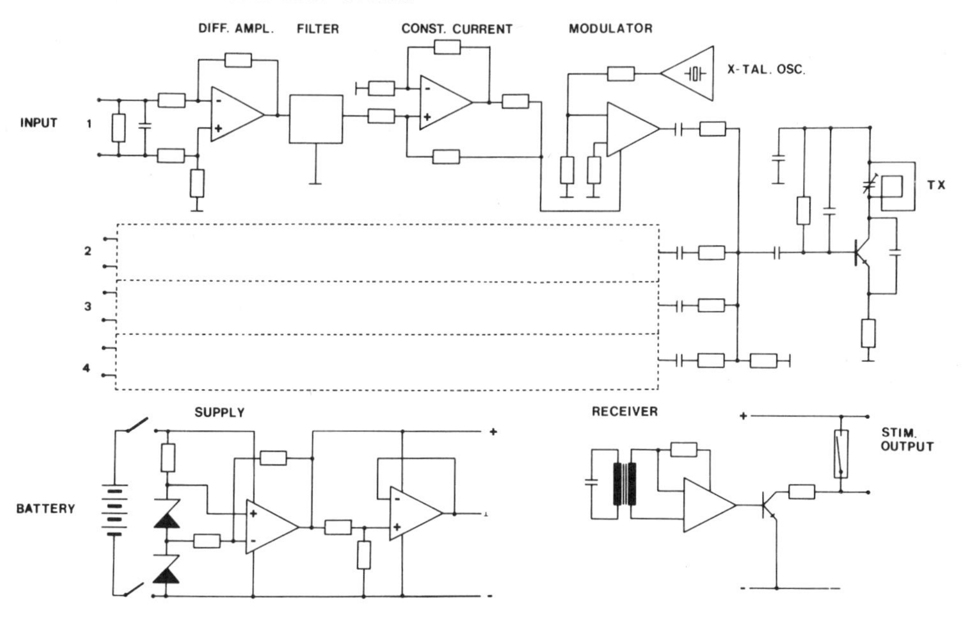

Fig. 1. Block diagram of the 4-channel telemetry system, the power supply and the stimulation receiver

Table I. Specifications of 4-channel telemetry system

TRANSMITTER		RECEIVER	
Frequency	104 Mc	Frequency range	100-108 Mc
Modulation	FM	Sensitivity	2.5μV/40dB/SN
Subcarrier freq.	10/12.5/16/19 Kc	IF bandwidth	200 Kc
Modulation	AM 50%	AF bandwidth	150 c
Input	4 sym. 47 KΩ	AF output	500 mV/50μV
AF range	1 - 150 c	AF range[1]	1-150 c/-3dB
Stim.receiver freq.	100 Kc	Selectivity[1]	-40dB
Stim. output	4 V sym.	Signal to noise ratio[1]	-45dB
Power supply	NiCd Akku 4.8 V		
Weight	50 g		

[1] at minimum antenna signal 15μV

The emitted signals are received by means of a conventional antenna and an FM receiver, demodulated and led into the amplifiers of a Grass polygraph. Faithful EEG and EMG tracings were recorded during long term experiments without any discernible crosstalk.

Rectangular pulses of 4 V, being variable in duration and frequency, can be delivered to a bipolar concentric stainless steel electrode located in a subcortical area by means of a stimulation receiver, working at 100 Kc. Since the frequency range of the transmitter is far from that of the pulse receiver, there is no interference between the stimulating and the recording telemetric circuits.

Preliminary experiments have shown the reliability of telemetric brain stimulation and recording over a long period of time. We are using now this technique in neuropsychopharmacological studies in cats, mainly in the study of drug action on sleep mechanisms. In order to correlate the EEG with the behaviour, the cats, housed in sound attenuated cages, are observed with tele-monitors without being disturbed by extraneous stimuli and the presence of the experimenter. In addition, a radar system allows the recording of the movements of the animal. In our opinion all this technical equipment , including telestimulation and telerecording, are necessary 1) to avoid many possible interferences with relevant data during long-term experimentation, particularly if automatic and computer analyses are to be used for storage and preprocessing of data, and 2) to collect as much information as possible about brain functions without restraining the animal.

References

BORBÉLY, A.A.; BAUMANN, I., and WASER, N.M.: Multichannel telemetry of physiological parameters (body temperature, ECG, EEG) in the rat. II. Applications in neuropharmacology; KIMMICH and VOS Biotelemetry, pp. 381 - 388 (Meander N.V., Leiden 1972).

HIMWICH, W.A.; KNAPP, F.M.; and STEINER, W.G.: Electrical activity of the dog's brain: Telemetry and direct wire recording; in HIMWICH and SCHADÉ Horizons in neuropsychopharmacology, Vol 16 of Progress in brain research, pp. 301 - 317 (Elsevier publishing company, Amsterdam 1965).

Biotelemetry II. 2nd Int. Symp., Davos 1974, pp. 179–181 (Karger, Basel 1974)

Telemetered Neurophysiological Responses in the Olfactory System During Habituation

Olfactory Bulb-Induced Waves-Single Neuron-Habituation

G. Verberne, F. Blom, J.N. de Boer and A.P.M. Schalken**
Laboratory of Animal Physiology, Amsterdam

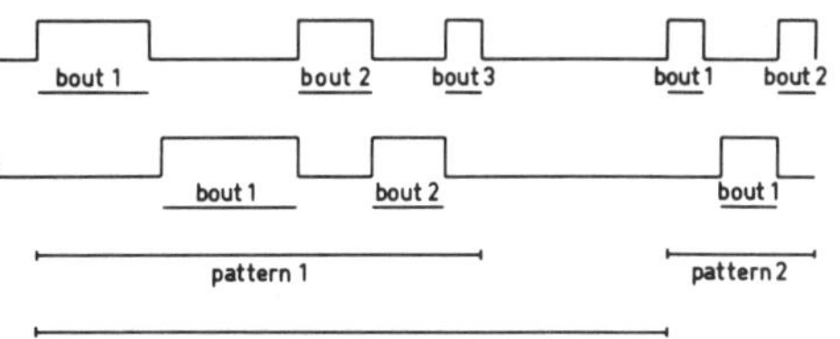

length sniffing of pattern 1 = bout 1 + bout 2 + bout 3
length sniffing of pattern 2 = bout 1 + bout 2
length flehmen of pattern 1 = bout 1 + bout 2
length flehmen of pattern 2 = bout 1

Fig. 1 An example of two trials of sniffing stimuli within a pattern of sniffing and flehmen bouts.

Introduction.

Cats and rabbits released in an observation arena containing pheromone carriers such as urine spend a great deal of time in olfactory exploration. Consequently they are able to select a stimulus, decide how long and how frequently they will explore it through sniffing and flehmen (Fig.1). A later return to the same scentsource results in a decrease in frequency and duration of sniffing and flehmen (7). The concomitant changes in the neurophysiological activity of the olfactory system is the subject of this study.

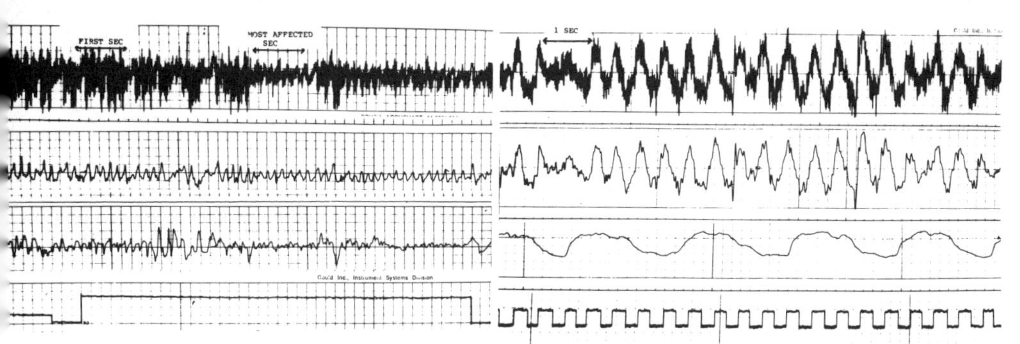

Fig.2 Slow respiratory and induced waves evoked by respiratory (left: rabbit M) and by artificial air puff stimuli (right: rabbit P7). Legend: from top to bottom, 1: induced waves, wide band, 2: respiratory waves shown by filtering the wide band signal < 15 Hz, 3: respiration, 4: left, sniffing period. right, air puff stimulus.
The 'most affected second' is defined as the second with the most pronounced decrease of amplitude. No response to sniffing is shown in the 'first second'. Amplitude of induced waves varies between 0.1 and 1 mV. Induced waves and respiration are recorded by telemetry, the former by a transmitter described earlier (6), the latter by a telemeter constructed by N.R.P. (Noordwijk, Holland).

* granted by Z.W.O. no. 88-38

Methods.

Implantation and telemetered recording techniques used throughout this investigation have been presented earlier(1,6).Qualitative descriptions of the typical behaviours have been previously described(2,7).Induced waves,a characteristic of the olfactory system,are considered 'evoked'responses' rather than 'spontaneous EEG'(Fig.2.). The mean amplitude per unit time was estimated by measuring the envelope of the amplitudes.Data collected from one cat and one rabbit are presented here but are consistent with results obtained from a number of other animals.Another cat showed plasticity in the activity of a single neuron in the prepyriform cortex throughout a 6 week period.Sequential differences in the parameters given below were tested by the sign-test.

Abbreviations:S:Sniffing Fl:Flehmen D:Duration Fr.:Frequency Fr.sp.:Frequency spikes A.ind.w.:mean amplitudes induced waves.

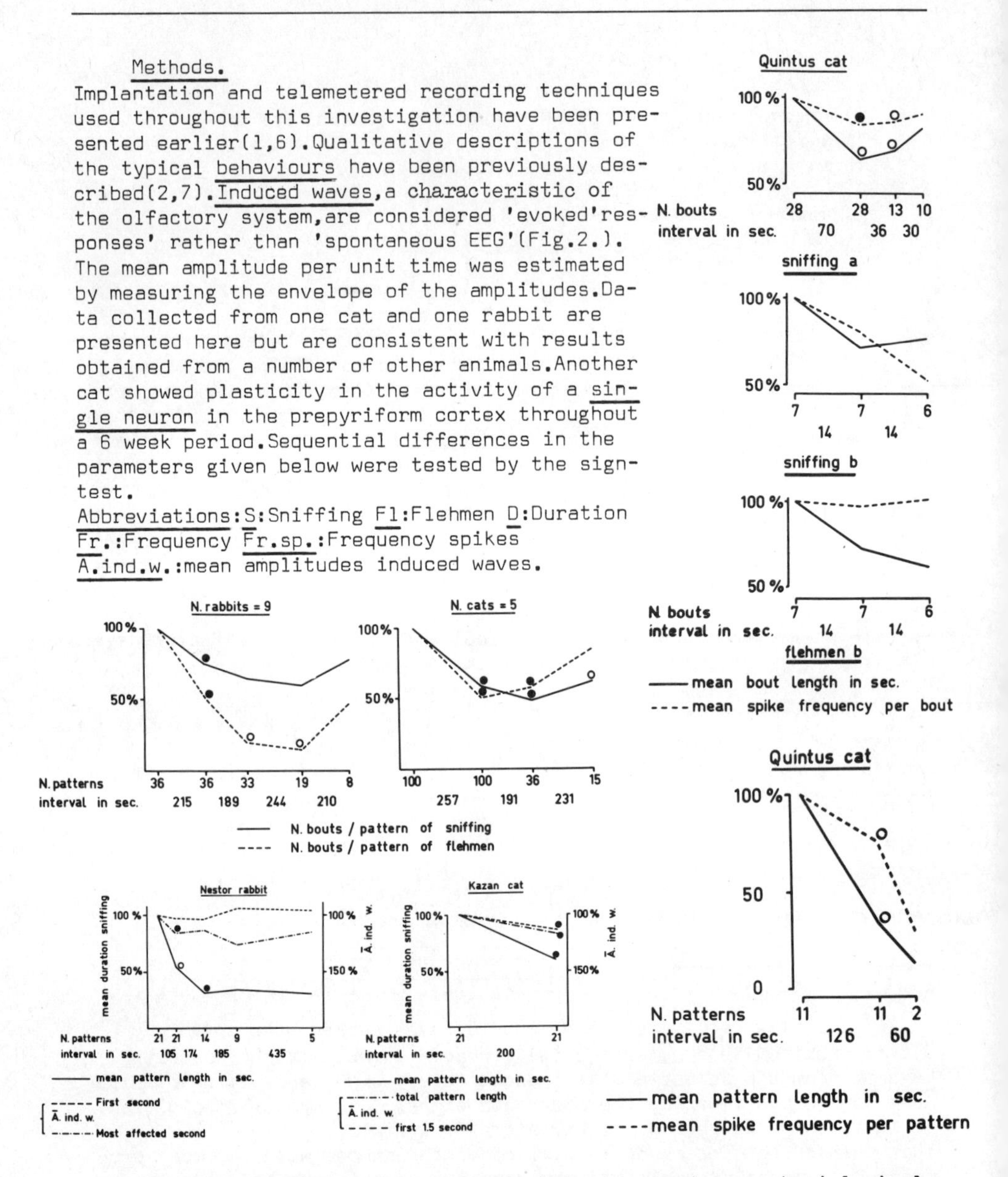

Fig.3 Concomitant habituation in behaviour and neurophysiological activity during repeated self-stimulation. The consecutive mean intervals are not equal but fail to show any trend. When values of D. and Fr. differ significantly from those of the first trial they are indicated by: o (P <.10) or ● (P <.05). The 7 (b) out of 28 (a) bouts of S. preceding Fl. are shown separately (cat Quintus).

Results and discussion. (Fig. 3)
1)Behaviour. D.and Fr.of bouts of S.and Fl.measured per pattern decreases with repetitive stimulation and is followed by a recovery; D.of S.bouts decreases within one pattern, those of Fl.do not, confirming earlier observations (2,7). Fl. wanes more rapidly than S.; this is, in general the rule for consummatory acts cf. orienting acts (4). No indication of cross habituation in S.and Fl.could be demonstrated for different scentsources.
2)The A.ind.w. decreases during S.(1,Fig.2) relative to that of the reference period (period of general arousal-cat, general olfactory exploration-rabbit). The diminuation of the decrease in A.ind.w. in the second trial was apparent for the total patternlength (cat) and for the most affected sec (rabbit) but less clear in the first sec of a pattern. Probably each time the animal returns he has to decide in the first sec whether to explore the scentsource more intensively or not. This interpretation is supported by the fact that habituation in S.does not occur after exploring objects without pheromones. The Fr.sp. habituates only during S., not during Fl.; consequently this neuron can rarely be considered as a 'non-specific attention unit'. We found no cross habituation for different scentsources, neither in the A.ind.w.nor in the Fr.sp.
3)The slopes of the relative attenuation of behavioural and neurophysiological variables in Fig.3 suggest that behavioural habituation proceeds more drastically than the neurophysiological one, confirming other investigations (5).
4)Although both simultaneously recorded behaviour and neurophysiological activity decrease during repetitive stimulation, pair-pair comparisons make it feasible that each habituation process occurs independently.
5)Inhibition of the induced waves, as well as habituation processes, are mediated by the Reticular Formation (3). Also the kind of sniffing affects the amplitude, as in separated bulbs, the decrease continues (2). We consider both possibilities the consequence of an arousal affect, caused by the significance of the stimulus; recognition leads to habituation. In either case, we assume we are dealing with the output of the information processing system.

References

1)BLACK-CLEWORTH,P. and VERBERNE,G.: Telemetered EEG and neuronal spike activity in olfactory bulb and amygdala in free moving rabbits; in Kimmich and Vos, Biotelemetry; pp. 317-325 (Meander, Leiden, 1972).
2)BLACK-CLEWORTH,P. and VERBERNE,G.: in preparation.
3)HERNÁNDES-PEÓN,R.: Neurophysiological correlates of habituation and other manifestations of plastic inhibition.
Electro enceph.clin. Neurophysiol.(Suppl.) 13: 101-114 (1960).
4)HINDE,R.A.:Behavioural habituation;in HORN and HINDE Short-term changes in neural activity and behaviour;pp.3-61 (Un.Pr.House, Cambridge, 1970).
5)HORN,G.: Physiological and psychological aspects of selective perception; in LEHRMAN, HINDE and SHAW. Advances in the study of behaviour; vol. II, pp 155-215 (Academic Press, New York and London, 1965).
6)MEYER,A.A.: A simple wide-band F.M. transmitter for telemetering biopotentials; Med. Biol. Eng. (in press).
7)VERBERNE,G.: Beobachtungen und Versuche über das Flehmen katzenartige Raubtiere;
Z.f.Tierpsychol. 27: 807-827 (1970).

Biotelemetry II. 2nd Int. Symp., Davos 1974, pp. 182–184 (Karger, Basel 1974)

EEG-Telemetry in the Rat: Selective Recording from 5 out of 12 Chronically Implanted Electrodes*

R. Moser, M. Däniker and A.A. Borbély

Institute of Pharmacology, University of Zürich, Zürich

A recently developed method (BORBÉLY, BAUMANN and WASER, 1972; BORBÉLY et al., 1973; KRAFT and VOEGELI, 1973) made it possible to record electrical brain and muscle potentials in the unrestrained rat by 4-channel telemetry. When applying the method it became evident that for several reasons it would be useful to record from more than 4 chronically implanted electrodes: (1) the choice of the reference electrode and of the 'active' electrodes from a large number of implanted electrodes would allow making a more extensive use of a single experimental animal; (2) due to the pulse-interval modulation technique of the transmitter, signal synchronization problems occur if an electrode picks up high voltage artifacts. It would be useful in this case to be able to disconnect the faulty electrode; (3) for several applications it is desirable to use transducers such as an accelerometer as an additional input source for the transmitter. These were the main reasons which prompted us to develop the method illustrated on Fig.1 which allows connecting any 5 out of 12 chronically implanted electrodes to the transmitter.

The implanted brain and muscle electrodes are soldered to 12 parallel metal strips on the bottom side of an epoxy resin socket (18 x 33 x 12 mm) which is fastened to the animal's skull with dental cement. The 5 input channels (4 'active' and 1 reference electrode) are connected on the bottom side of the transmitter to 5 metal strips which come to lie perpendicularly to those of the socket. 5 jumpers (gold-plated beryllium bronce) soldered to the bottom side of the transmitter can be placed so as to contact any electrode with any channel of the transmitter. For recording, the transmitter is placed into the socket and held in position by a tightly fitting cover (Fig.2). The edges of the socket and cover are rounded off to prevent cage mates from gnawing on the transmitter assembly. The transmitter can be easily removed from the

* This study was supported by the Swiss National Science Foundation, grants nr. 3.8790.72 and 3.212.73.

animal's head for changing the battery or for contacting different electrodes. Optionally, a miniaturized strain gauge accelerometer (BORBÉLY et al.,1973; DÄNIKER,1973) can be positioned on the top side of the transmitter and connected to one of its input channels. This makes it possible to obtain quantitative records of the animal's motor activity simultaneously with the brain and muscle potentials.

References:

BORBÉLY,A.A., BAUMANN,I. and WASER,N.M.: Multichannel telemetry of physiological parameters (body temperature, ECG,EEG) in the rat. II. Applications in neuropharmacology; in KIMMICH and VOS Biotelemetry; pp. 381-388(Meander, Leiden 1972).

BORBÉLY,A.A.,DÄNIKER,M.,MOSER,R. and WASER,P.G.: Multiparameter telemetry in neuropharmacological research. AGEN-Mitteilungen 15: 29-34 (1973).

DÄNIKER,M.H.: Subminiature accelerometer for radiotelemetric recording of motor activity in the rat; in Proceedings of the 10th Conference on Medical and Biological Engineering, p.61 (Dresden,1973).

KRAFT,W. and VOEGELI,F.: 4-Kanal Miniatursender zur Uebertragung des Elektroencephalogramms von Kleintieren. AGEN-Mitteilungen 15: 19-24 (1973).

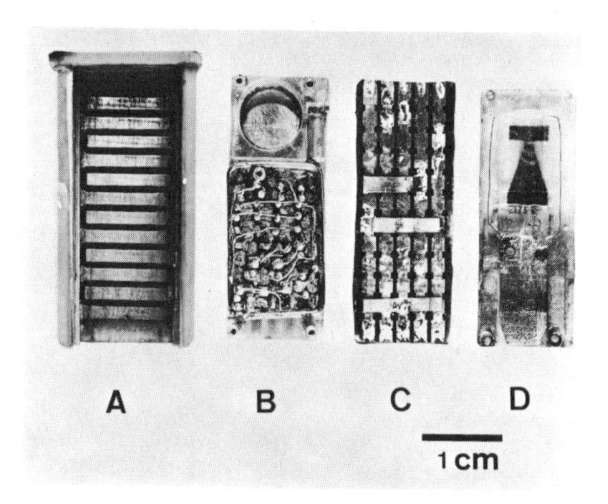

Fig.1 Transmitter assembly. A: Top view of socket. B: Top view of transmitter with battery case. C: Bottom side of transmitter with jumpers soldered to input channels. D: Strain gauge accelerometer to be positioned on top of the transmitter.

Fig.2 Rat with transmitter assembly.

Biotelemetry II. 2nd Int. Symp., Davos 1974, pp. 185–187 (Karger, Basel 1974)

Behavioral and Methodological Factors Influencing Heart Rate of Freely Moving Rats

Béla Bohus

Rudolf Magnus Institute for Pharmacology, Medical Faculty, University of Utrecht, Utrecht

Relations and interactions between behavioral and physiological responses have long provided a major source for the study of emotional behavior. Heart rate measurement has been used frequently to obtain data on autonomic responses accompanying emotional behavior. Changes in heart rate have been viewed as relevant alterations to behavioral processes involved in emotion, motivation, learning or somatic activity accompanying behavior. However, due to contradictions reported in the literature it is not yet possible to develop a uniform view about the "meaning" of heart rate changes during emotional behavior. A comparison of data obtained with various recording techniques suggests that methodological factors might be one cause of contradictory results. Accordingly, the present experiments were aimed to explore 1) whether the heart rate changes as measured with the aid of telemetry or through wire-leads are comparable in various behavioral situations and 2) whether the mode of recording affects the behavior of the rat.

Methods

Male albino rats of an inbred Wistar strain, weighing 160-170 g, were used. The electrocardiogram (ECG) of the rats was recorded bipolary from electrodes implanted under the skin. When wire-leads were used, a noiseless light-weight cable was attached to the electrodes. Free movement of the rats was allowed by the counterbalanced cable. The cable was connected with a Grass A.C. Preamplifier and the ECG was recorded on a Grass Mod. 79B polygraph. The data were stored for analysis on magnetic tape using a Philips Ana-log 7 taperecorder.

Telemetric recording of the ECG was made by using a light-weight miniature FM transmitter (Mod. SNR 102F, Smith and Nephew Res., Ltd.) with a carrier frequency of 102.2 MHz. The transmitter was mounted on a 10 mm wide harness which was secured round the thorax. The signals were received by a biotelemetry receiver (Narco-Bio Systems Inc., Mod. FM-1100-7). The filtered and preamplified ECG signals were displayed on a polygraph and stored on tape.

Heart rate was determined by measuring the interbeat intervals with the aid of a PDP 8/1 computer. The mean interbeat interval was computed for selected time-periods depending upon the data acquisition in various behavioral test situations.

Results and Discussion

Heart rate of rats housed in single cages was measured in a sound-proof experimental room while the animals were left in their home cage. Each observation period lasted 15 min daily for 5 days. ECG records were taken at 30 sec intervals for 15 sec. Mean interbeat intervals were computed for 5-min blocks.

Heart rate of the rat attached to the recording devices through wire-leads did not change significantly either within or between the experimental sessions. In contrast, a significant decrease within the sessions ($p < 0.05$) and between the sessions ($p < 0.05$) was observed in rats from which heart rate was obtained by telemetry. This clearly indicates that the heart rate of rats bearing a transmitter shows an adaptation to handling both within and between the sessions. Rats attached to wire-leads failed to display a similar adaptation.

Heart rate responses during exposure to a novel environment were studied in a circular open-field (2). Behavioral measures included ambulation, rearing and grooming scores recorded for 15 min daily for 5 days. Heart rate was measured at 30 sec intervals during motionless periods and mean interbeat intervals were computed for 5-min blocks.

Although the mean heart rate of rats attached to wire-leads showed a tendency to decrease both within and between the sessions, analysis of variance did not indicate significant changes. In contrast, a significant decrease in heart rate was observed in rats bearing a transmitter both within ($p < 0.01$) and between ($p < 0.05$) the sessions, i.e. rats bearing a transmitter displayed a rapid adaptation to the novel environment. It is worth mentioning that the behavior in the open-field was not different in the two groups of rats.

Heart rate changes accompanying a learned emotional behavior were also studied in a "step-through" passive avoidance situation (1). The rats were trained to enter a large dark compartment from an elevated, illuminated platform located in front of the large box and connected to it through a guillotine door. After 4 pretraining trials an unavoidable footshock (0.25 or 0.5 mA A.C. current for 1 sec) was given in the large compartment through the grid floor of the box. Retention of the passive avoidance response was tested 24,48 and 72 hours after the shock trial by measuring the latency to reenter the shock compartment. Heart rate was continuously recorded during the last preshock trial and during the retention tests when the rat was staying on the elevated platform. Passive avoidance behavior was accompanied by a decrease in heart rate when compared to the pre-shock values. Both the shock intensity ($p < 0.01$) and the retention interval ($p < 0.05$) had a significant effect on the degree of bradycardy. These main effects were present whether ECG was recorded through wire-leads or with the aid of telemetry. However, the absolute heart rate of rats with wire-leads was always higher than that of rats bearing a transmitter. Passive avoidance latencies of the rats with wire-leads were also always longer.

In conclusion, these comparative studies on the heart rate of freely moving rats as recorded with the aid of telemetry or through wire-leads attached to ECG electrodes clearly indicate that apart from various behavioral factors the mode of recording may have very pronounced influences on this autonomic measure. Accordingly, methodological factors should also be considered when the "meaning" of autonomic responses accompanying emotional behavior is the subject of investigation. From a physiological point of view, the use of telemetry in this type of experiments is superior to record heart rate changes.

References

1. ADER, R., WEIJNEN, J.A.W.M. and MOLEMAN, P.: Retention of a passive avoidance response as a function of the intensity and duration of electric shock. Psychon. Sci. 26: 125-128 (1972).

2. BROADHURST, P.L.: Experiments in psychogenetics: applications of biometrical genetics to the inheritance of behaviour. In EYSENCK Experiments in Personality: Vol. 1. Psychogenetics and Psychopharmacology, p. 30 (Routledge & Kegan Paul, London 1960).

Session VIII

Patient Monitoring – Clinical Telemetry

Chairmen: *H.P. Kimmich, T. Furukawa and J.A. Vos*

Biotelemetry II. 2nd Int. Symp., Davos 1974, pp. 190–195 (Karger, Basel 1974)

Clinical Telemetry and Patient Monitoring

Present Situation and Proposals for Possible Improvements

H.P. Kimmich

Department of Physiology, University of Nijmegen, Nijmegen

There are several fields in health care where clinical telemetry and patient monitoring is incorporated. However it should be stated clearly that in the clinical sphere both biotelemetry and patient monitoring play only a limited role as compared to its capabilities. This obviously has many reasons, some of them with distinct origin in the medical and physiological field or the daily clinical practice. In such a situation it should, however, be well considered if the methods and equipment available are possibly inadequate to cope with the special situation they should be used for. It is likely that many papers will put forward this question and deal with the problem in the near future. Let us consider first the situation as it stands today (fig. 1).

Field of Application / Applied Method	Clinical Surveillance	Function Tests	Rehabilitation	Remote Diagnosis	Mobile Clin. Emergency Systems	Biological Research	Sport and Work
Patient Monitoring (Cont. measurement via wires)	X	X	-	-	-	X	(X)
Wired Telemetry (incorp. multiplex techniques)	-	(X)	-	(X)	-	(X)	(X)
Wireless Telemetry							
Short Distance 0 to 10 m	(X)	X	(X)	-	-	X	(X)
Medium Distance 10 to 500 m	(X)	X	X	-	-	X	X
Long Distance up to some km	-	-	(X)	(X)	X	(X)	(X)
Combined Telemetry	-	-	-	-	-	(X)	-
Storage Telemetry	(X)	-	X	-	-	(X)	X
Interface and Transducer Problems							
Subject - Equipment	X	(X)	(X)	X	X	(X)	-
Equipment - Man	X	-	X	-	X	-	-

Figure 1: Present situation of application of biotelemetry and patient monitoring in clinical practice, biological research, and sport and work.

It may be noticed that telemetry plays only a minor role in routine clinical work. Especially striking, however, is the rare use of wired telemetry and the almost entire absence of incorporation of combined telemetry (wired telemetry in combination with short range telemetry). It is not the suitability that has prevented application of these methods in health care systems, since they combine many of the advantages of medium range biotelemetry such as earth free operation and flexibility in display and processing with the absence of disadvantages known from radiotelemetry (frequent replacement of batteries, restrictions of frequencies and/or number of channels, possible RF interference, etc.). The major advantage of radiotelemetry (freedom of movement of the patient within certain limits) is either not required in the hospital or in the major cases sufficiently accomplished by the use of combined telemetry (freedom of movement in,and in the vicinity of,the patient's bed, e.g. within the room). Obviously a remaining disadvantage of wired or combined telemetry is the necessity of laying a cable. With combined telemetry a single cable is necessary per room carrying the information of one or more parameters of several patients in one room as well as the information of all other patients connected to the system. Display and processing equipment can be connected to the system at any place, e.g. at a central remote position (data processing and storage, patient administration), at the remote surveillance ends, or at the patient ends. Processing and/or display is possible at each point of an individual patient, a group of patients with simultaneous or selected display (automatic or manual), or of all patients with display of one or more selected parameters.

Such a system also allows execution of some of the more disputed problems such as inquiry of patient data at the patient end out of the central processing and storage unit, reminding the patient or nurse (e.g. for use of medicine) and execution of decisions taken by next generation processing equipment, since the combined telemetry is a true two-way system.

Displacement of the patient to another room or the operating theater does not call for dis- and reconnection of the patient to the system. Contrary to radiotelemetry, however, the biological information is then transmitted on a different channel, which calls for new patient identification. In an elaborate system this can be done directly from the patient end. The fact that the system primarily identifies the room and not directly the patient seems to be a disadvantage only at first sight. Also,with omission of correct patient identification an automatic alarm always identifies correctly the place where something is wrong.

Simultaneous medium range radiotelemetry using the same subcarrier (e.g. for rehabilitation purposes) is compatible with the combined telemetry system.

A block diagram of a two-way combined telemetry system as it could be realized simply and relatively cheaply is shown in figure 2. Notice that the frame of such a system is relatively simple and that the additional equipment is identical for all patients. The system measures up to 4 parameters (DC to 100 Hz) of up to 160 patients. Such a system is compatible with auxiliary equipment (transducers, data pro-

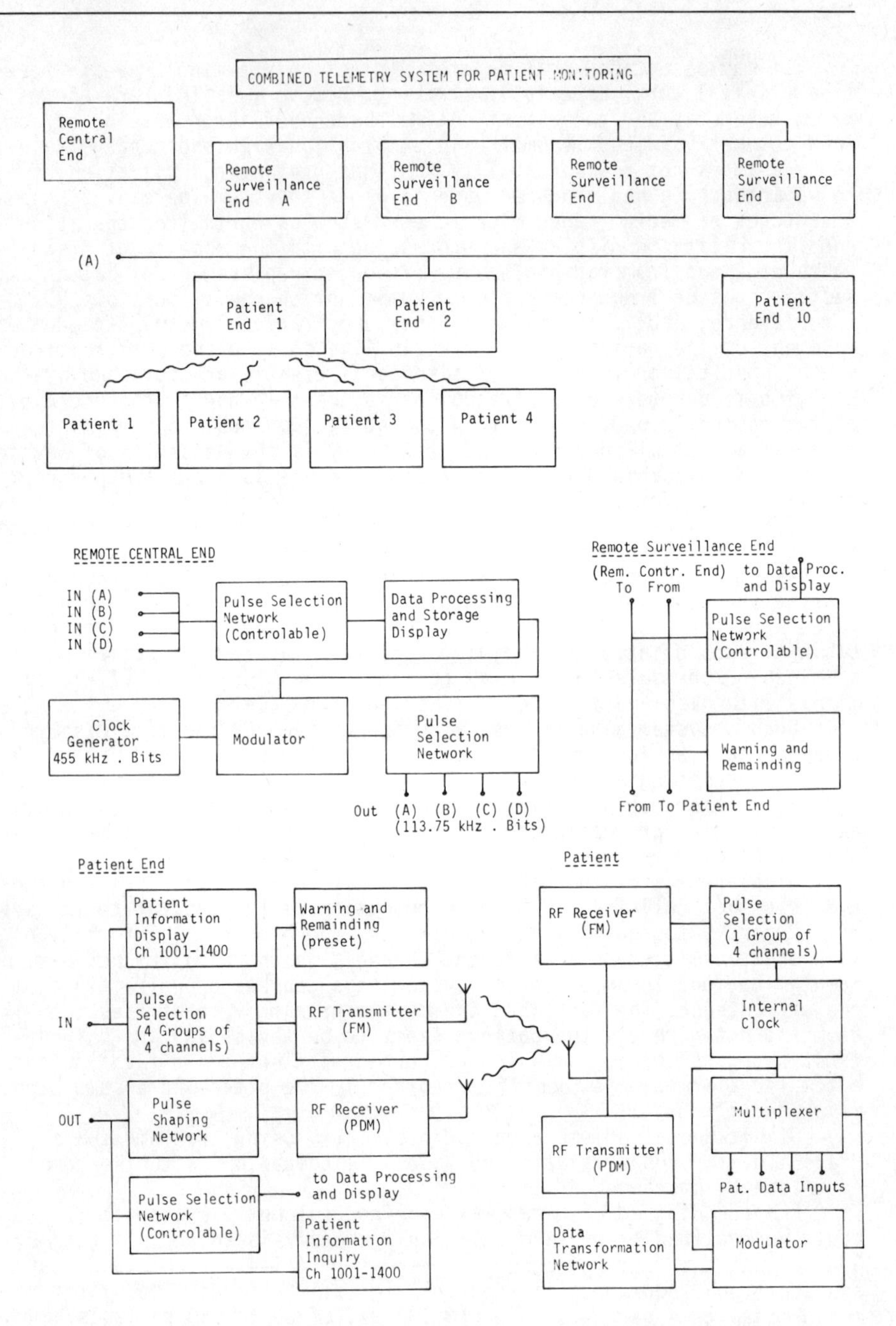

Figure 2: Block diagram of a combined telemetry system.

cessing and display equipment, etc.) as used at present in most hospitals. It merely reduces the number of interconnecting wires, introducing at the same time a number of advantages mentioned earlier.

Working of the two-way combined telemetry system:

The remote central end unit is connected with a two-way coaxial cable to four groups of up to 10 patient end units each. One to several remote surveillance end units can be connected in each of the connecting lines A, B, C, and D (fig. 2). Connection from the patient end units to the patients is realized with short range radiotelemetry.

The pulses of a 455 kHz clock generator for PAM and PDM or a 455 kHz times the number of bits generator for PCM are (after possible modulation with warning, reminding, and patient data) selected so that channels 1, 5, 9, 13, etc are connected to patient ends 1 to 10, channels 2, 6, 10, 14, etc. to patient ends 11 to 20, etc. At each patient end the corresponding 4 groups of four channels are selected, and then frequency modulate an RF carrier. The information, radiated with less than 1mW, is received by 4 patient units. The corresponding 4 channels are selected, recorded with new patient parameters, delayed, and after PDModulation of an RF carrier,retransmitted. By incorporating PDM, total transmission time is less than 1/350, thus the power of the patient unit can be extremely small,allowing operation of several months from a single small battery. The information of the four patients is received by the corresponding patient end unit consecutively and,after pulse reshaping,available at each point of the system for data processing, display, storage, etc.

While the first 1000 channels are reserved for continuous patient parameter transmission the remaining 400 channels may be used for patient administration and information transmission. From the "pulse-channel-patient" relation scheme (fig. 3) it may be noticed that two of the group synchronization"pulses" generally omitted may be used as speech channels (SCh) in both directions if the corresponding 16 pulses in the"patient administration and information band" (pulses 1001-1400) are also reserved for this purpose (e.g. for music transmission to the patients).

PULSE TRAIN (A)

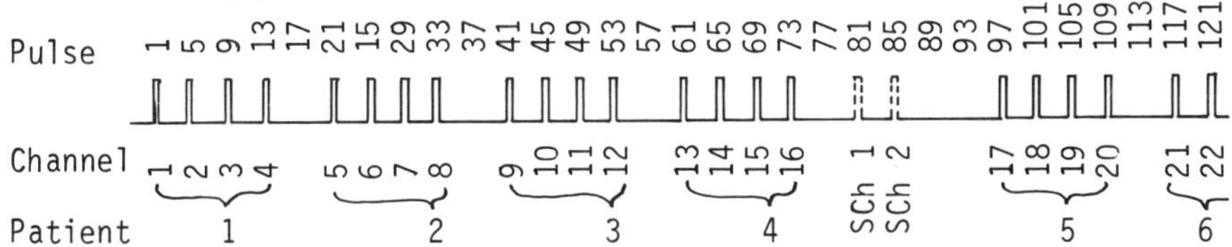

Figure 3: Pulse-channel-patient relationship scheme of a combined telemetry system.

The realization of a combined telemetry system for clinical use would be easy and large-scale production would bring many advantages to the hospital but is also feasible at a reasonable price on a small-scale basis

Unfortunately there exists, besides the transmission problem, the well-known interface and transducer problem, subject-equipment (fig. 1) which is a real problem in more or less all fields of clinical monitoring. Contrary to the transmission problem the interface and transducer problem is difficult and cannot be solved generally. New methods, specially designed for correct and artefact-free accession of biological parameters in the moving patient are likely to be successful in solving part of the mentioned problems. However, besides that it is difficult to find new methods (most of the methodological work concentrates on the adaption or improvement of old principles), such new methods often show new, up-to-now unknown problems. A good illustration is the dynamic oxygen uptake ($\dot{V}O_2$) measurements (e.g. Kimmich and Kreuzer, 1972). When measuring $\dot{V}O_2$ by continuously integrating the product of the respiratory flow and O_2 concentration, the response time of the flow and O_2 concentration curve, have an important and not merely time-dependent influence on the accuracy of the O_2 measurement. For a proper design of such equipment it is important to know the share of such an artefact on the total inaccuracy. A qualitative answer is obtained on a theoretical basis while the qualitative answer is obtained from a model based on the theoretical considerations and as many measurements as possible confirming the correctness of the model at least under

REST

ION

$\ldots PO_{2_{insp}}\ e^{-t/\tau_2}$

$\ldots \frac{\cdot f \cdot t}{\ldots} = \dot{V}_{Max} \cdot t/t_2$

$\ldots C_{O_{2_{insp}}} \cdot \dot{V}_{Max} \int_0^{t_2} e^{-t/\tau_2} \cdot t/t_2\ dt$

ON

$\ldots PO_{2_{insp}} \cdot f(t)$

$(t + t_3)/t_2$ (for $0 < t < t_2 - t_3$)

$(2 - (t + t_3)/t_2)$ (for $t > t_2 - t_3$)

$\ldots O_{2_{insp}} \cdot \dot{V}_{Max} \left[\int_0^{t_2 - t_3} f(t)\ (t + t_3)/t_2\ dt + \int_{t_2 - t_3}^{\infty} f(t)(2 - (t + t_3)/t_2)\ dt \right]$

$f(t) = \left(\frac{\tau_2}{\tau_2 - \tau_1} e^{-t/\tau_1} + \frac{\tau_2}{\tau_1 - \tau_2} e^{-t/\tau_2} \right)$

$t_2 = 15/(8/100\ \dot{V}_{Max} + 6.4)$

$t_3 = t_2 \cdot 18/\dot{V}_{Max}$

WORK

INSPIRATION

$\Delta PO_2 = \Delta PO_{2_{insp}}\ e^{-t/\tau_2}$

$\dot{V} = \dot{V}_{Max} \cdot t/t_0$ (for $0 \leq t \leq t_0$)

$\dot{V} = \dot{V}_{Max}$ (for $t > t_0$)

$$\Delta \dot{V}O_2 = \Delta C_{O_{2_{insp}}} \cdot \dot{V}_{Max} \left[\int_0^{t_0} e^{-t/\tau_2}\ t/t_0\ dt + \int_{t_0}^{\infty} e^{-t/\tau_2}\ dt \right]$$

EXPIRATION

$\Delta PO_2 = \Delta PO_{2_{insp}} \cdot f(t)$

$\dot{V} = \dot{V}_{Max}\ (t + t_1)/t_0$ (for $0 < t < t_0 - t_1$)

$\dot{V} = \dot{V}_{Max}$ (for $t > t_0 - t_1$)

$$\Delta \dot{V}O_2 = \Delta C_{O_{2_{insp}}} \cdot \dot{V}_{Max} \left[\int_0^{t_0 - t_1} f(t)\ (t + t_1)/t_0\ dt + \int_{t_0 - t_1}^{\infty} f(t)\ dt \right]$$

where $f(t) = \left(\frac{\tau_2}{\tau_2 - \tau_1} e^{-t/\tau_1} + \frac{\tau_2}{\tau_1 - \tau_2} e^{-t/\tau_2} \right)$

$t_1 = 3.6/\dot{V}_{Max}$

$t_0 = 0.2$ sec

ERROR IN OXYGEN UPTAKE MEASUREMENT

$$E = (\Delta \dot{V}O_{2_{insp}} - \Delta \dot{V}O_{2_{exp}})/\dot{V}O_2$$

Figure 4: Equations for error calculations of dynamic O_2 measurement

PO_2 = partial pressure of oxygen in (mm Hg)

C_{O_2} = oxygen concentration in (%)

$\dot{V}$ = flow in (ℓ/min)

$\dot{V}O_2$ = oxygen uptake in (ℓ/min)

f = respiratory rate in (1/min)

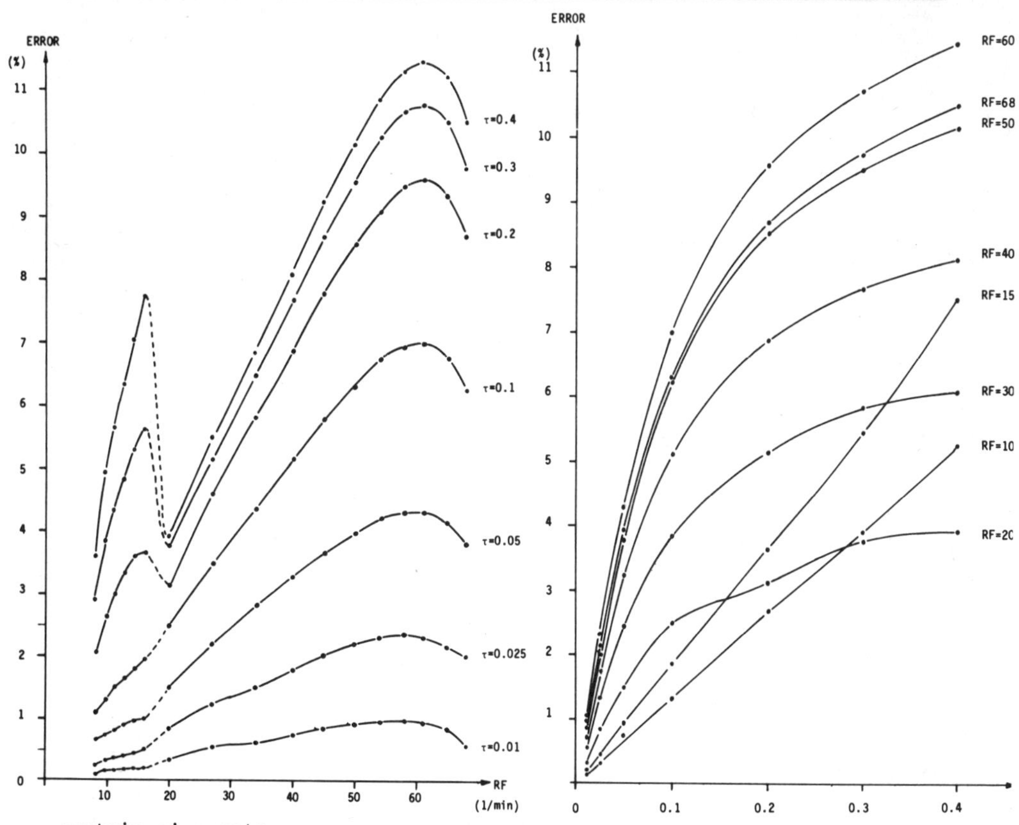

certain circumstances.

Calculations of the errors to be expected in dynamic oxygen uptake measurements as a function of the time constant (τ_2) of the oxygen measuring system (the flow measuring system is very fast) have been done by approximating the respiratory flow and PO_2 curve mathematically (fig. 4). Description of the PO_2 curve is accurately possible under practically all conditions, while the flow curve approximation is done on a worst case basis by using a triangular waveform during exercise (with elevated respiratory frequency) and a trapezoid waveform at rest or moderate exercise. The result is shown graphically in figure 5.

Conclusions

At present,telemetry and patient monitoring play only a limited role in clinical routine work as compared with its possibilities. This may have several reasons. The main reason (problems of interface of patient-equipment) has been well known for some time. Solution of this problem will take much time due to its difficulty and complexity. Alternative transmission equipment (e.g. combined telemetry) can stimulate increased application of patient monitoring.

References:

KIMMICH,H.P. and KREUZER,F.(1972) TELEMETRIC DETERMINATION OF OXYGEN UPTAKE BY MAN DURING EXERCISE. Int. J. of Biomedical

Biotelemetry II. 2nd Int. Symp., Davos 1974, pp. 196–198 (Karger, Basel 1974)

Uses of Telemetry in Health Care

Health Care Telemetry

John Hanley

Space Biology Laboratory, Department of Psychiatry and
Department of Computer Science, University of California, Los Angeles, Calif.

Local Telemetry Epilepsy of the temporal lobe, so-called psychomotor epilepsy, is the commonest form of epilepsy in the USA. It appears to be the consequence of hypoxia at birth and head injury in later life. Unfortunately, it is often very resistant to anticonvulsant medication and as many as 40% of patients (more in some series) are not helped, and they are at the mercy of their seizures. A considerable number of these patients can be helped if it can be determined that the seizure had a unilateral locus. Spontaneous bilateral seizures are beyond help at this time. Attempts to record the EEG by conventional hardwire means relies on the serendipitous capture of a seizure during a scheduled recording: not only may this not happen even after many weeks of recording, it is a well known clinical phenomena that admission to hospital often is accompanied by marked decrease of seizures even to the point of absence. Thus an extremely costly hospitalization up to 6 weeks in duration may fail to provide the neurosurgeon with a diagnosis. A telemetered recording of a temporal lobe seizure with a unilateral focus enabled a neurosurgeon to perform a unilateral temporal lobectomy, and this patient has not been seizure-free for several years.

The utilization of such telemetry systems which free the patient from being hard-wired to conventional bulky machines has resulted in increased yield of diagnostically important seizures, and reduced hospital stay from the previous six weeks to 10 days. The systems at present are frequency-domain multiplexing devices which use Inter-Range-Instrumentation Group (IRIG) standard center frequencies of the proportional bandwidth type. These are standard for NATO countries as well as the USA. The system has been extensively described elsewhere and will be but briefly recounted here: the amplified EEG signals frequency-modulate subcarrier oscillators. The outputs of the oscillators are then linearly summed, and this summed signal amplitude-modulates a crystal controlled transmitter. The main carrier signal at 90 MHz is detected by a receiver and its output goes to subcarrier discriminators which match the original center frequencies of the VCOs, converting the voltages back to frequencies. The signals are then written out on a conventional EEG machine or stored on FM magnetic tape. This proportional bandwidth frequency domain system has served well for 10 years but will soon be replaced by a time domain pulse position system.

Remote Telemetry Virtually every country in the world, ranging from great to small, has difficulties in providing high quality medical care to remote, often indigent areas. Modern telecommunication systems can help to put the resources of the modern medical center, university or community based, at the service of the rural community, or city ghetto. In order to economically further transmit the locally acquired signal, it is necessary to match the signal to the existing communication medium by optimal modulation techniques. In most instances the channel will be voice grade telephone lines and the modulations will be in the frequency domain. Modulation in time division multiplexing schemes in general will require too high a data sampling rate for the usual voice grade line and the installation of special high-rate lines would cost enormous amounts of money. Proper exploitation of existing lines requires knowledge of their bandwidths so that the number of center frequencies that can be used with proper guard band protection may be determined. In the USA, most domestic lines have a bandwidth from 400 Hz - 3 KHz. The use of the proportional bandwidth center frequencies determined by the IRIG US military group after World War II allows eight channels of data to be sent over the voice-grade lines. Two methods are used to accomplish this. In one, the number of the hospital is simply dialled and the audio tones inductively coupled to the phone by holding the handset of the phone close to the loudspeaker of the receiver. This simple method works well when the ambient noise is low. The second method is the use of an amateur radio phone patch which allows switching from data to voice. Examples of clinically important signals transmitted from a small hospital which lacks the sophisticated monitoring capability of our Laboratory include EEG signals from patients on artificial renal hemodialysis and in critical care units. Existing phone systems can also be used for routine testing of devices such as pacemakers, and for checking on the efficacy of anti-arrhythmic drugs in the case of heart disease, and anti-convulsant drugs in the case of epilepsy. In certain circumstances, the patient will not require aid in the data transmission process: in other circumstances judicious use of other personnel such as the public nurse will be necessary.

Another grave and challenging problem outside of the domain of the hospital that appears to exist in the Western hemisphere, including the United States, is that of intelligent children who have serious difficulties in learning to read. Such children can be extremely disruptive because of the attendant frustrations and some authorities relate aggression in the classroom to the disorder of developmental reading dyslexia. It is possible to study these children by telemetry and relay the signal by voice grade telephone to the laboratory for analysis. Computer analysis of normal time domain EEGs from these children shows them to generate unusual frequencies from the left parieto-occipital area which differentiate them from normal children with no reading problems.

With the aid of the U.S. Air Force Office of Scientific Research, it has been possible to extend the study to young adults. Initial findings indicate that severe dyslexics have EEG spectra that resemble the children. They show poor development of narrow band alpha, and have peaks in the theta range. On the other hand, dyslexics that appear to have benefited from remedial approaches appear to generate a well-developed alpha peak but retain some of the theta activity that is more characteristic of dyslexic subjects. Because of the expense of the computation, however, efforts are now being made to see if a special-purpose device can be utilized in the doctor's office.

Where telephone systems are limited and terrain difficult for line-of-sight requirements, satellites can be utilized. Our Laboratory has demonstrated inter-continental transmission of EEG and EKG signals between Sweden and the USA. One advantage of such communication is that medical research requiring powerful computers which a small country may not have direct access to is made possible by linking the small country to the large one: data was transmitted from Dr.David Ingvars' Laboratory in Lund, Sweden, and received at the Space Biology Laboratory, Los Angeles, USA, enabling the data to be analyzed by our IBM 360-91 general purpose digital computer.

Finally, although there are many other areas of interest in Biotelemetry, there is an area of emergency care in which telemetry can play a role in addition to the transmission of the medically important information. That is, in the speedy transportation of the emergency case to the hospital. In the case of myocardial infarction, from 40-70% of the victims in the U.S. die in the first hour. Clearly, the Soviet Skoroya system with its specialized cardiac ambulances is the best way so far to meet this problem. However, much of the world is not organized like this and the problem of a speedy transfer by emergency vehicle to a treatment center, often a dangerous ride through dense urban traffic, remains. I propose two methods to aid this process: 1) Equip emergency vehicles with a device that could control the traffic lights on route (one could use a version of the siren) and 2) make standard equipment in all vehicles a visual warning that an ambulance is en route and consequently they should pull to one side.

Acknowledgements This work was supported in part by Air Force Office of Scientific Research Contract # F44620-73-C-0070.

Biotelemetry II. 2nd Int. Symp., Davos 1974, pp. 199–201 (Karger, Basel 1974)

Telecommunications in Health Care Systems*

M. Bracale and G. Scarpetta
Elettronica Applicata, Facoltà di Ingegneria, Università di Napoli, Napoli

In the modern Health Care Systems transmission and processing of biomedical data are becoming steadily more important. This depends upon the possibilities offered by the technology involving also the immediate use of computers.

In Health Systems many fields are interested in the use of telemetry which can be implemented with various communication links (wireless, hardwired, optical).

In Health Care Systems there are many departments: some for prevention, some for diagnosis, some for therapy and rehabilitation.

In these departments the information fluxes have different purposes, but in an integrated system they can interfere.

Therefore, it is convenient to establish standards for implementary modular structures for coupling with others not only to record and display the signals, but also to compute on-line or off-line biomedical data for different purpose. This transmission can be either wired or wireless; in Table I some examples of two fields are summarized.

TABLE I

Medicine	Transmission	
	WIRED	WIRELESS
Preventive		Sport-medicine Occupational-medicine
Diagnosis	Link with specialized centers	Ambulance and specialized centers
Therapy	-Local hospitals (Coronary Centers C.C.U. Intensive Care Units I.U.C.) -Surgical rooms (specialized centers)	-Foetal monitoring -Intensive care units -Surgical rooms -Patients in transit
Rehabilitation	patient at home	patient at hospital

There are two main fields of applications for biotelemetry:

1) in distant transmission for diagnosis on subjects outside the health area;
2) in transmission for monitoring on patients in hospitals or generally in the health area.

---

* The present work was partially supported by the Consiglio Nazionale delle Ricerche (C.N.R.) Roma, Italy.

Examples of the first field of application can be found in information transmission for distance diagnosis (between zone hospital and regional hospitals, or between ambulances or the patient's house and information or computing centers). In this case a wired connection is required.

Examples of the second field of application are in:

a) intensive care units
b) coronary care units
c) surgical rooms
d) nuclear therapy units
e) insulated patients
f) foetal monitoring units

In these cases, wired and wireless connections are requested, although the last one is always prefered for reasons of compactness as compared with traditional monitoring and especially, for safety considerations. It is also necessary in rehabilitation departments where the patients perform their normal living functions.

From the above examples, it follows that the systems are generally multichannel-systems for wired or wireless connections. Therefore, modular structures are required. Based on these considerations, our interest is in studying implementing and comparing modems for the two kinds of transmission. Because of the present prices of fast analog-digital conversion equipment , the choice is for analog transmission, with which it is also possible to use public telephone line without coming into conflict with the regulations as one would in the case of digital transmission. An other choice to be made is between AM and FM systems.

Keeping in mind the bandwidth of the telephone line, the spectra of the main biological signals and a minimum number of channels (such as three for the VECG applications) an FM system can only be implemented with very low modulation indices. Therefore FM and AM systems become practically equivalent regarding to S/N ratio.

Our group has implemented AM and FM systems for collecting biomedical data in the computing center from local hospitals.

In this paper a comparision between these two implemented systems is reported.

Fig.1 shows the responses of the best and the worst of the four channels in the AM system.

In Fig.2 those of the FM system are given, and in Fig.3 the overall bandwidth is drawn.

T A B L E II	AM SYSTEM
Channels	4+1 (synchroniser)
Bandwidth for each signal channel	0.2 - 100 Hz
Carrier frequencies	1200,1800,2400,3000 Hz
Input voltage level for 80% modulation	IV
Power level at input of telephone line	0 dBm
Output voltage level	0.6 V_{pp}
Crosstalk attenuation	42 dB
Carrier frequency for synchronizing pulse	5 kHz
Width of synchronizing pulse	50 ms

	TABLE III	FM SYSTEM
bandwidth	(channels 1,3)	0 ÷ 50 c/s
bandwidth	(channels 2,4)	0 ÷ 70 c/s
modulation index	(channels 1,3)	2
modulation index	(channels 2,4)	1
overall bandwidth		400 ÷ 3000 Hz
carrier frequencies		800;1300;1800,2300
S/N		> 40 dB
crosstalk		> 40 dB

In Table II and III the main performances of the two systems are summarized.

From measurements and results reported which are in accordance to the theoretical previous, the two systems are equivalent especially for S/N ratio considerations.

We think that the final choice must be carried out on the base of constructive simplisity.

Looking to the telephone applications and bearing in mind the corresponding bandwidth, an FM system is convenient for the transmission of those biomedical signals with spectra localized in the very low frequency zone (temperature, pH measurements and so on); on the contrary transmission of those signals with larger spectra forces one to use extremely low modulation indexes giving thus worse S/N ratios as compared an equivalent AM system.

The authors thank Prof.F.Cappuccini for useful suggestions.

References (See the paper "A Four-Channel FM-System for Radio and Telephone Transmission"; same authors, same section)

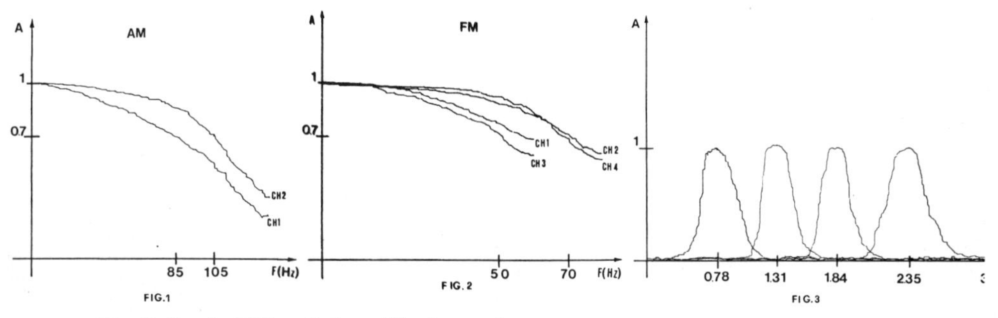

Fig.1 Bandwidth of two AM channels
Fig.2 Bandwidth of four FM channels
Fig.3 Overall bandwidth of the FM system

Biotelemetry II. 2nd Int. Symp., Davos 1974, pp. 202–204 (Karger, Basel 1974)

A Multichannel Biotelemetry Transmitter Utilizing a PCM Subcarrier

Thomas B. Fryer and Richard M. Westbrook
NASA, Ames Research Center, Moffett Field, Calif.

Multiple channel telemetry systems for biomedical applications in the past have usually used FM subcarrier oscillators because of their advanced development in avionics. In recent years aviation, and especially space and satellite telemetry systems, have most commonly been PCM because of its high accuracy, but its size has been too large for biomedical applications. With the advent of new integrated circuits, and especially low power complimentary MOS (CMOS), the implementation of miniature PCM transmitters is now feasible. Making use of the currently available integrated circuits, a miniature (less than 90cm^3) PCM telemetry system operating from one miniature 9-volt battery and drawing less than 5ma has been constructed.

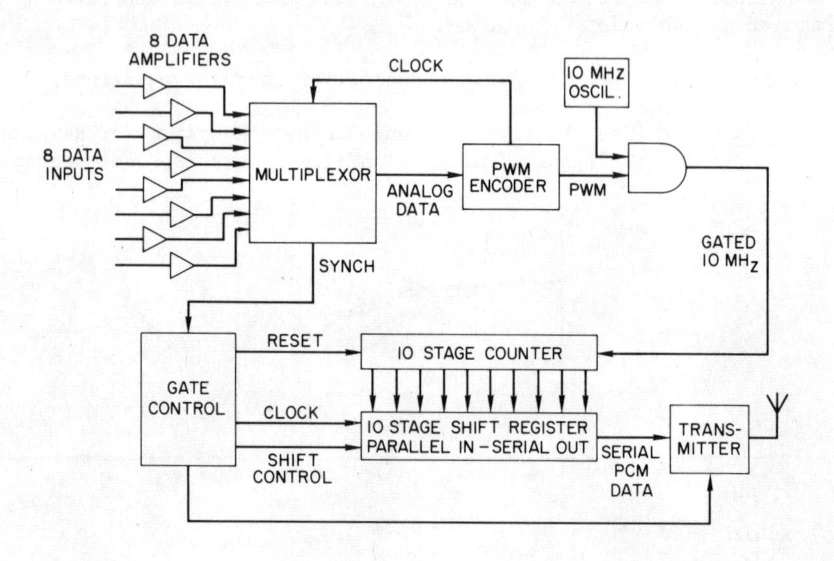

Fig. 1 A block diagram of the Multichannel PCM Telemetry System

A block diagram of the system is shown in Fig. 1. A separate preamplifier is used for each data channel to provide appropriate gain to bring the transducer signals up to a maximum of ±1 volt. The analog data is then multiplexed and converted into a Pulse Width (PWM) format. To obtain the PWM signal, the analog data and a ramp generator signal are inputed to a comparator, as shown in Fig. 2. Channel 0 is blanked for frame sync identification and channel 1 has a zero input to provide an accurate zero reference. This is then followed by eight data words to form one frame. The PWM signal is converted to Pulse Code Modulation (PCM) by gating a 10MHz oscillator on and off with the PWM signal and then counting the number of 10MHz cycles. Typical waveforms are shown in Fig. 2. The counted cycles at the end of each word are parallel-shifted to a shift register and the counter is reset to zero, ready to accept the next word. While the next word is being counted, the previous word is serial-shifted out of the shift register as a PCM signal. Both the PCM signal and a sync signal are transmitted out by a simple one-transistor FM transmitter.

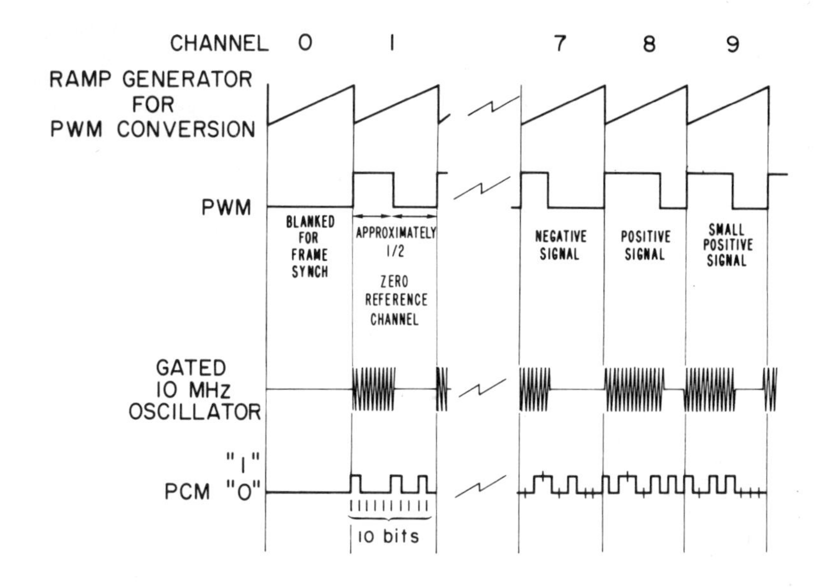

Fig. 2 Typical waveforms generated in the system.

A frame rate of 1000 samples per second is used and with the 10 words per frame, each word is 100 usec. duration. The PWM is restricted to a range of 10 usec. to 90 usec. to allow gating time in the counter and shift register. This, then, converts to a 100 to 900 count with the 10MHz oscillator. The 10-bit binary coding (1024) provides an accuracy of ± 0.1% with the intention that adequate resolution is obtained even with a data scale of 1/10 the system capability. Since ± 1% accuracy is adequate for most biological data, this provides a wide dynamic range without the need for scale changes. The 1000 samples per second per channel provides a theoretical 500Hz upper frequency limit, but 250 to 300Hz is a more practical limit, depending upon the type of filter used. Subcommutation can be used to provide additional lower frequency data channels.

The implementation of the system is achieved using RCA COS/MOS devices to provide extremely low power operation. A miniature 9-volt radio battery provides ± 4.5v for the system. Although the COS/MOS will operate with as low as a 3-volt supply, higher voltages are needed for 10MHz operation and for the linear IC's. The ± 4.5v exceeds the minimum device requirements so that reliable operation is achieved, while at the same time is low enough to minimize any shock hazards that might occur. Total current requirements, including the preamplifiers, is about 5ma. The RF stage uses about 1ma of the 5.

The use of a PCM telemetry signal not only provides accurate data transmission, but requires a minimum of RF bandwidth. A 10-bit code at 10,000 words per second is 100,000 bits per second and this can easily be accommodated in a 200KHz FM band. Both the 88-108MHz and the 180-216MHz RF bands have been used. The serial format that results from multiplexing the input data allows the use of a single channel recorder with its cost savings over a multichannel tape recorder. An additional benefit of the PCM encoding is that data is in digital format and is directly suitable for computer use.

Biotelemetry II. 2nd Int. Symp., Davos 1974, pp. 205–207 (Karger, Basel 1974)

A Four-Channel FM-System for Radio and Telephone Transmission

M. Bracale, G. Scarpetta and M. Buonomano
Elettronica Applicata, Facoltà di Ingegneria, Università di Napoli, Napoli

In modern Health Systems there are numerous departments with links for the transmission of biomedical data.

Sometimes the transmission area is well defined and concerns very short distances; in other instances, the links are in less welldefined areas and with variable distances.

Examples of the first kind are the departments inside an hospital area (Intensive Care Unit; Coronary Care Unit; Rehabilitation Centers; Foetal Care Unit). Examples of the second kind are transmissions within the processing center and other laboratories or the patient's house.

Therefore, some cases could be served by radiotelemetry, others also by public telephone links.

It could be convenient to have picking-up and transmitting systems with modulation structures utilizable in both ways.

The present paper describes a multichannel FM and frequency division modulation system to be linked with the public telephone line or a radio transmitter.

Access to the telephone line is implemented by two electromagnetic couplers, compatible with the present regulations.

Description of the System: The system implements a frequency modulation by varying the slope of a triangular waveform according to the amplitude of the analog input (Fig.1).

The conversion amplitude-frequency is obtained through a level inverter, driven by a voltage comparator and an integrator. Through a nonlinear network the triangular waveform becomes an FM sinusoid.

In Fig.2 the demodulator is shown. It consists of a zero crossing detector and a system of Butterworth filters.

The main characteristic of the system that was implemented is the wide range of linearity, obtained in a very simple way (Fig.3).

The specifications are given in Table I.

bandwidth	(channels 1,3)	0 ÷ 50 c/s
bandwidth	(channels 2,4)	0 ÷ 70 c/s
modulation index	(channels 1,3)	2
modulation index	(channels 2,4)	1
complete bandwidth		400 ÷ 3000 Hz
carrier frequencies		800;1300;1800;2300
S/N		> 40 dB
crosstalk		> 40 dB

Telephone coupling is obtained by using the magnetic fluxes dispersed from the earphone. The apparatus is equalized by considering the bandwidth of the couplers (Fig.4).

The most significant part of the apparatus has been described. For wireless transmission we used an HF-FM modulator and an Astro Communication Receiver.

The authors thank Prof.F.Cappuccini for useful suggestions.

References:

BRACALE,M., and RUGGIERO,F.: Applicazioni Mediche della Biotelemetria. L'Automazione nell'Assistenza Sanitaria, ANIPLA-Milano, 193-203 (1971).

BRACALE,M.: Considerazioni sulla biotelemetria: realizzazione di una unità monocanale.
Rapporto FUB-Roma (1970).

BRACALE,M.: Teletrasmissione dei segnali biologici.
Note Recensioni e Notizie dell'Istituto Superiore PT-Roma 6, (1970).

BRACALE,M., and SANTORO,M.: Trasmissione via radio di segnali elettromiografici.
55° Congresso di Ortopedia e Traumatologia, SIOT-Napoli (1971).

BRACALE,M., RUGGIERO,F., and SCARPETTA,G.: Trasmissione multicanale di segnali biologici.
Primo Convegno Mostra di Bioingegneria, FAST-Milano (1972).

BRACALE,M., RUGGIERO,F.: Multichannel telephone system for biomedical applications.
Med. and Biolog. Eng., 10, 688-691 (1971).

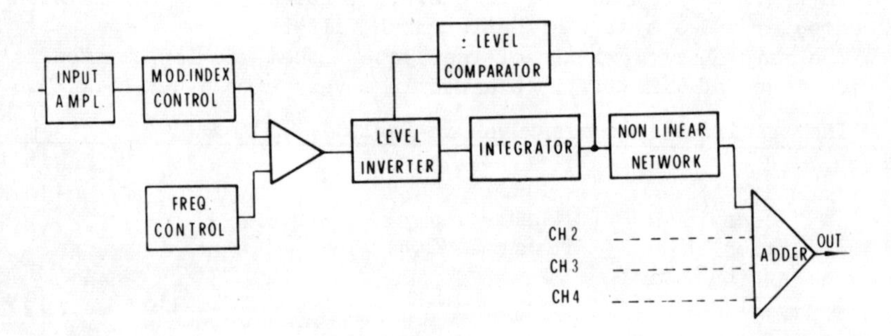

Fig.1 Block diagram of the transmitter

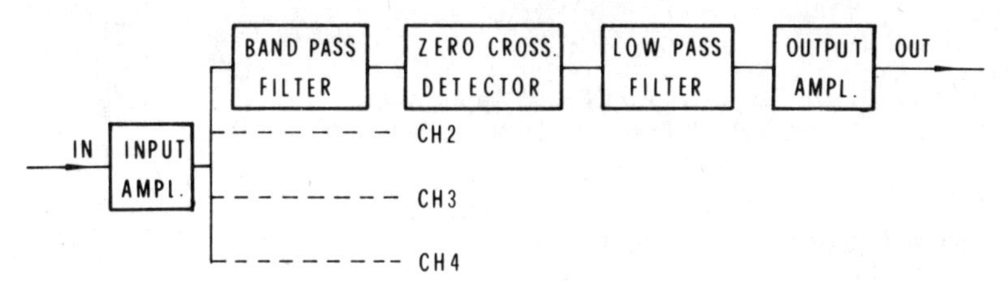

Fig.2 Block diagram of the receiver

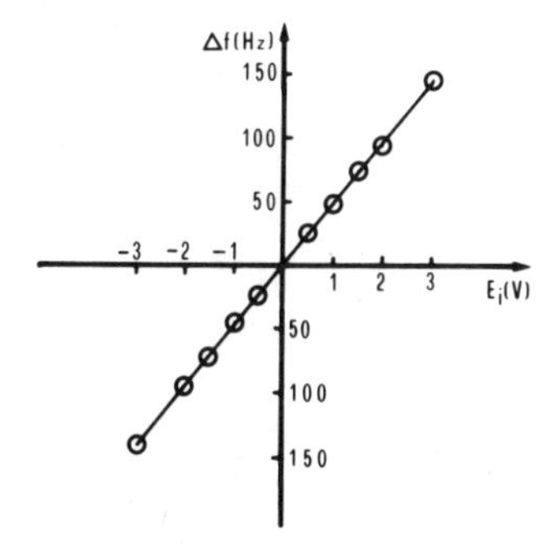

Fig.3 Modulation characteristic of the system

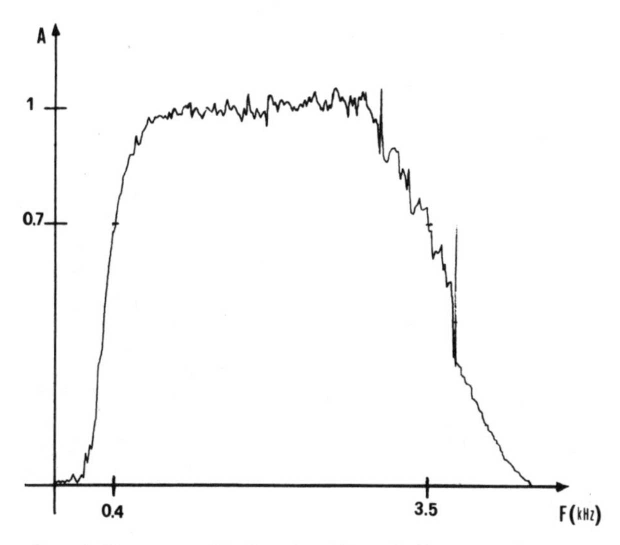

Fig.4 The overall bandwidth of the coupler

Biotelemetry II. 2nd Int. Symp., Davos 1974, pp. 208–210 (Karger, Basel 1974)

Radio or Wired Telemetry for Clinical Applications?

Patient Safety Standards Improved by Wired Telemetry

C. Weller

Clinical Research Centre, Harrow, Middlesex

The rapid increase in continuous surveillance or patient monitoring has led to a reconsideration of safety standards for electro-medical equipment. It is reasonably easy to devise means by which a single instrument can be rendered adequately safe, but it is more difficult to ensure that leakage currents under fault conditions remain within safe levels when several instruments, possibly connected to different power outlets are used simultaneously on the same patient. One method of ensuring that assemblies of instruments remain safe with minimal reliance on the skill or vigilance of the staff is to use 'earth free' or isolated input stages to all instruments.

The exceptionally high degree of isolation afforded by radio telemetry coupled with freedom of movement might appear to make this the method of first-choice. In fact, radiotelemetry has had very little impact on routine clinical monitoring. There are several reasons for this, among the more significant are:-

i) restrictions on the frequency bands available.
ii) expensive equipment requiring servicing facilities not normally available in hospitals,
iii) storage and replacement of batteries,
iv) radio frequency interference.

The disadvantages listed above only apply to the use of radio, it can be shown that although the degree of isolation is not so high, there are many advantages of using the encoding and power conserving techniques originally developed for radiotelemetry but over a hard wire. As an example the design of an isolated ECG monitor is given.

The conventional way of isolating a patient from potentially dangerous voltages is to link the transducer or in the case of an electrocardiogram, the electrode leads, to a remote bedside unit; isolation by means of transformers or opto-isolators and subsequent signal processing then takes place in the bedside unit.

The proposed method uses low power electronics and encoding techniques to shift the more sensitive and interference prone parts of the signal processing from the bedside unit to a small patient borne unit (50x85x20 mm) which is linked to the monitor or display by a lightweight twin lead. This lead which is not screened and very flexible is isolated at both ends, which means that if it becomes abraided and touches either the live side of the power line or ground the patient is not endangered.

This contrasts markedly with conventional systems where long leads linking the patient to a bedside unit are prone to pick up interference and contact dangerous potentials.

As well as supplying power for the electronics and supporting the patients encoded ECG, the twin lead is used to send signals to the patient; in the case of the ECG monitor this facility is used for calibration. It could equally well be used for control purposes such as electrical stimulation or drug infusion.

Although wired telemetry does not allow the same freedom of movement as radiotelemetry, very satisfactory results can be obtained in exercise work with treadmills and bicycle ergometers; the lightweight flexible lead of virtually unlimited length is not an encumbrance in these applications.

A block diagram of the system is shown in figure I; the electronics are separated into two parts; the 'Patient End' and 'Remote End'.

At the remote end power for the patient end circuits is derived from a 25kHz oscillator. This oscillator may be amplitude modulated by a 2.5kHz tone generator. The power is coupled to the patient end by transformers at each end of the twin lead. Because the power requirement for the patient end electronics is only a few milliwatts these transformers are designed not for efficiency but for high insulation resistance and low interwinding capacitance. The leakage current is less than 1µA when 240Vrms are applied to the patient. The 25kHz power is rectified and smoothed; the 2.5kHz modulation that appears at this stage is used to initiate a calibrator that appears in series with the electrodes. The power supply is then regulated to supply the remaining circuits.

The ECG signal after suitable amplification is converted into very short duration pulses (1µS) of variable rate which are sent to the remote end for decoding. These narrow pulses are easily separated from the power supply by pulse transformers and capacitors.

The present single channel instrument may easily be modified for multichannel use. The narrow encoding pulses are ideal for time division multiplexing applications and as there are no bandwidth restrictions very high data rates are possible. Similarly more tone generators may be added so that frequency division multiplexing may be used to send additional information or control signals to the patient end.

During the past decade there have been many new instruments developed for diagnostic applications; it seems likely that the next generation of instruments will not only record data but will make decisions and take action based on the recorded data. Wired telemetry is a safe and reliable way of transferring data to or from patients.

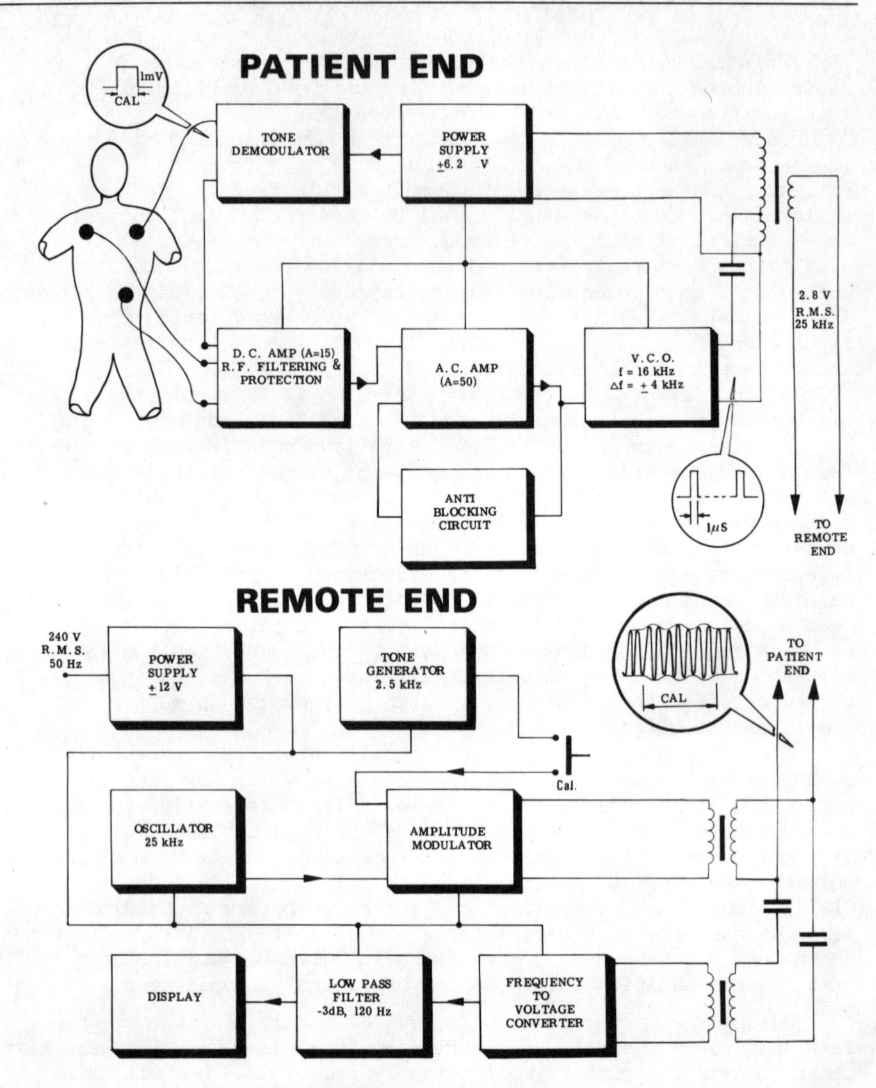

FIGURE 1. Schematic Diagram.

Biotelemetry II. 2nd Int. Symp., Davos 1974, pp. 211–215 (Karger, Basel 1974)

Long-Term Telemetry of Direct Arterial Blood Pressure in Unrestricted Hypertensives

B. Krönig, K. Dufey, P. Reinhardt, J. Jahnecke and H.P. Wolff
I. Medizinische Klinik und Poliklinik, Johannes-Gutenberg-Universität, Mainz

1. Introduction: It is well known that the indirect method of blood pressure recording (RIVA-ROCCI/KOROTKOFF) is likely to give incorrect results when applied to persons under physical activity (e.g. ANSCHÜTZ und DRUBE, 1952; KARLEFORS, NILSEN, and WESTLING, 1966). Because of the higher velocity of the pulse wave under physical activity the systolic value is indirectly usually estimated too high whereas the diastolic pressure is determined too low, sometimes it may not be discernable at all. Most of the information about the every-day blood pressure behaviour is thus restricted to variations observed under resting conditions. Evidence of how much the "resting blood pressure profile" is altered by the blood pressure reaction to every-day physical activity, especially in hypertensive patients, is lacking. In our opinion, thorough investigations of this problem were necessary, since they will give an idea of the actual blood pressure load on the cardiovascular system under every-day conditions.
Two methods for continuous direct intra-arterial blood pressure recording in unrestricted men have been described so far: 1) BACHMANN and THEBIS (1967) applied a large-scaled ÖDMAN-LEDIN-catheter transfemorally having the pressure transducer fixed in front of the chest; flushing of the catheter had to be done intermittently at 20 to 30 minute-intervals. The data was transmitted by radiotelemetry leaving the possibility of stretching or condensing the blood pressure curve at the receiving unit (see also contribution ZERZAWY et al., Session VI/1). 2) BEVAN, HONOUR, and STOTT (1969) inserted a thin nylon catheter into the brachial artery having the transducer and the pump for continuous perfusion put together in a box strapped onto the chest at heart level. Primarily a recording galvanometer and camera were placed in another small box to be carried in a jacket pocket. Thus, a more comprehensive blood pressure curve was obtained without the possibility of beat-to-beat differentiation, but registering blood pressure for at least 24 hours. In the last two years the authors (LITTLER et al., 1972) adapted a small portable magnetic tape (see contribution McKINNON, Session III/2) which made it possible to replay the records at various speeds and thus differentiate beat-to-beat afterwards.

2. Method and Material: Especially for more extensive and long-lasting studies in a greater number of unrestricted hypertensive patients we developed the method of the micro-catheter radio blood pressure telemetry (for preliminary data see KRÖNIG et al., 1972) which in some ways may be understood as a combination of the two methods mentioned:

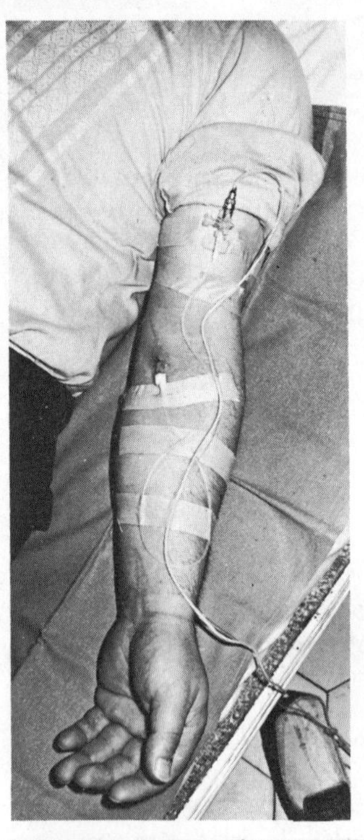

Figure 1: Micro-catheter radio blood pressure telemetry; equipment adapted to the patient's arm.

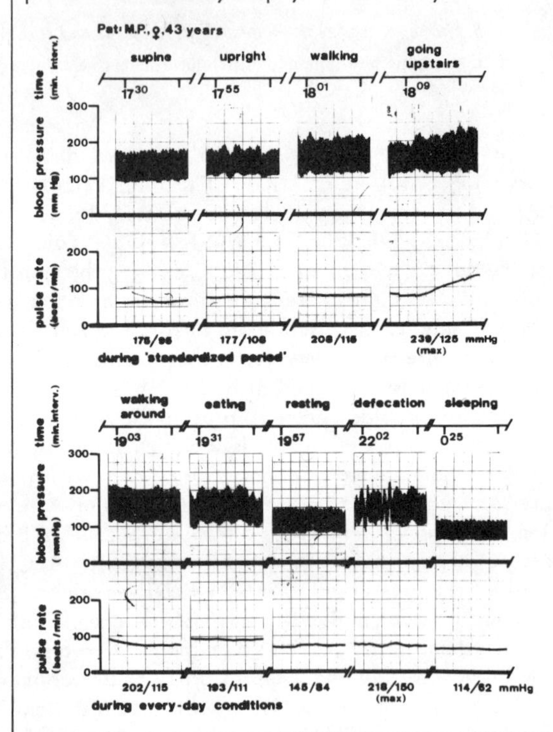

Figure 2: Cut sections from an original chart of a long-term blood pressure telemetry in a woman with essential hypertension; blood pressure variability to physical activity.

1. After local anaesthesia the brachial artery is located by percutaneous palpation and a small plastic catheter (Intranüle, Fa.VYGON) measuring 1.2 mm for the outer, and 0.8 mm for the inner diameter, is inserted into the brachial artery for the whole length of 120 mm. This percutaneous route of application does not cause much more discomfort to the patient than an intravenous injection. The plastic catheter leaves movements of the patient's arm unrestricted.
2. The catheter is connected to a conducting plastic tube which ends at the miniature transducer (SP 37, Fa.STATHAM) fixed at the patient's upper arm (Figure 1).
3. The attached cable being paralleled by the small plastic flushing tube is lead from the transducer side to the patient's equipment, which is packed together in an easy-to-handle bag weighing approximately 2.3 kg.
4. The bag contains an accumulator operated pressure modulator and an one-channel transmitter (Fa.HELLIGE) with a bandwidth of 170 Hz; the FM/FM-system operates within 1.300 Hz for the pressure modulation between 0 and 300 mm Hg (second carrier frequency); the first carrier frequency is 151.01 MHz. The power consumption of the modulator including the transducer is

guaranteed by Ni-Cd-cells lasting for at least 12 hours, then recharging has to take place for another ten to twelve hours. During this time the blood pressure recording is continued; the transmitter is operated with exchangeable batteries lasting for 12 hours and then being recharged externally. Further, a plastic bottle with sterile saline containing 10.000 units of heparin per liter surrounded by a pressure cuff (Fa.FENWAL) is packed into the bag and connected to the flushing tube by interposing an automatic valve (Intraflo, Fa.SORENSEN). By inflating the cuff to a pressure above 300 mm Hg the whole system can be flushed continuously with a microvolume of 3 or 6 ml per hour, which is sufficient to keep the catheter free of blood.

5. The receiver set, placed in the laboratory, consists of a high-frequency receiver, the demodulator and instruments indicating the actual systolic and diastolic blood pressure as well as the pulse-rate; further, blood pressure and pulse-rate are recorded on an original chart (paper-speed 0.5 mm/sec) and parallel on an analog magnetic tape, leaving the possibility of stretching or condensing the registered data.
6. The continuously recorded blood pressures are interpreted from the original chart by estimating systolic and diastolic values geometrically over a period from 30 to 60 seconds. Single values, as for example the maximum of the blood pressure rise to physical activity, are taken as the mean of three consecutive pulse pressures, which - at the paper speed mentioned - are still separately distinguishable. Computer analysis of the data, by analog-digital converting of the magnetic tape information, is in progress.

During the whole long-term blood pressure registration the patient is asked to behave as close as possible to every-day conditions, especially to the extent of physical activity, e.g. walking and going upstairs at "his rate". According to the transmitter capacity he is allowed to move around in the hospital area as far as 600 m from the receiver set. Continuous measurements are carried out for at least 24 hours, usually the patient's every-day blood pressure is observed for 2 1/2 days, including three consecutive nights. In order to get comparable results of the blood pressure reaction to every-day physical activity, all patients have to accomplish so-called standardized periods (20 min recumbency, 10 min active orthostasis, 10 min walking, and going upstairs two flights) at four to five different times per day. As illustrated in Figure 2 with cut sections from an original chart of a long-term blood pressure telemetry in a 43-year-old woman with essential hypertension, the blood pressure changes registered under observation during the performance of a standardized period (upper part of the figure) are of the same extent as those occuring under every-day activities (lower part), except, of course, the lowest value during sleep.

3. Application and Results: The method has been applied in over 200 long-term studies (24 to 60 hours duration) of blood pressure behaviour in unrestricted hypertensive patients. The procedure was well tolerated in all cases, thromb-embolic complications did not occur.

Out of the numerous results obtained so far (KRÖNIG et al., 1973, 1974) one aspect may be mentioned here, demonstrating the range of applicability of the method in investigating the every-day blood pressure reaction in different groups of hypertensive patients at various times of the day (Table I): As already may be

Table I: Mean blood pressure values under every-day conditions at different times of the day, separated by the severity of hypertension (WHO-stages); values obtained during long-term radio blood pressure telemetry in unrestricted hypertensive patients.

mean values mm Hg (syst./diast.)		supine	upright	walking	going upstairs
WHO I (n=11)	morn.	141.2/ 86.2	142.6/ 92.1	151.5/ 88.8	180.9/ 97.8
	noon	138.7/ 82.3	142.2/ 89.3	157.5/ 89.3	185.5/ 98.7
	even.	143.5/ 85.1	151.1/ 93.1	162.5/ 90.1	186.1/ 96.6
WHO II (n=52)	morn.	162.8/ 90.0	160.0/101.4	174.8/100.0	202.0/106.0
	noon	160.1/ 90.0	160.6/100.3	175.2/ 97.7	206.7/104.1
	even.	164.8/ 93.2	166.4/102.0	179.3/100.4	210.4/106.6
WHO III (n=16)	morn.	203.8/111.5	194.9/118.9	214.4/121.3	245.7/123.9
	noon	211.3/113.3	203.7/121.5	221.1/118.2	254.7/125.6
	even.	212.0/112.9	213.4/124.6	224.7/119.0	258.3/129.8

seen in the mean values, the blood pressure reaction to every-day physical activity, as walking and going upstairs, is of almost equal extent in the three different hypertensive patient groups; thus, even in advanced hypertension (WHO III), the same blood pressure reaction is observed as in mild hypertension (WHO I). Compared to the preceding value in recumbency, the blood pressure reaction to every-day physical activity is augumented as the day passes by; the highest values are recorded in the evening. Further, the systolic blood pressure alteration in active orthostasis (upright) is different in the morning, at noon, and in the evening; so, blood pressures in orthostasis should not be compared to one another, unless they are taken at the same time of the day.

4. Discussion: In our opinion, the described method is a useful tool to get more precise and extensive information about the variability of the every-day blood pressure. This applies especially to hypertensive patients, since there is little known about their blood pressure reaction to every-day physical activity, being an important contributor to the pressure load of the induvidual cardiovascular system.
Compared to our method, the transfemoral application of a large-scaled catheter and the necessity for intermittent flushing (BACHMANN and THEBIS, 1967) limits the applicability of this method for serial long-lasting investigations in hypertensive patients. On the other hand, the method of BEVAN, HONOUR, and STOTT (1969) - though applicable for long-term blood pressure recording - fails to give a beat-to-beat differentiation of the blood pressure curve simultaneously with the measurement, which is important especially while recording hypertensive patients.

5. Summary and Conclusion: To permit continuous direct recording of blood pressure in a greater series of unrestricted hypertensive patients, the micro-catheter radio blood pressure telemetry was developed and applied by modification

of the methods of BACHMANN and THEBIS (1967) and BEVAN, HONOUR, and STOTT (1969). A micro-catheter is applied percutaneously into the brachial artery and connected to a miniature transducer; telemetric equipment and the unit for continuous flushing are packed together in an easy-to-handle bag. The method has proven effective in over 200 long-term studies (24 to 60 hours duration); serious complications did not occur. It is a valuable tool for studying blood pressure behaviour in unrestricted hypertensive patients.

6. References:

ANSCHÜTZ, F. und DRUBE, H.C.: Über den Fehler der auskultatorischen Blutdruckmessung nach RIVA-ROCCI/KOROTKOFF bei Kreislaufumstellungen. Verh.dtsch.Ges.Kreisl.-Forsch. 20, 278 (1954)
BACHMANN, K. und THEBIS, J.: Die drahtlose Übertragung kontinuierlicher, direkter Blutdruckmessungen. Z.Kreisl.-Forsch. 56, 188 (1967)
BEVAN, A.T., HONOUR, A.J., and STOTT, F.D.: Direct arterial pressure recording in unrestricted men. Clin.Sci. 36, 329 (1969)
KARLEFORS, T., NILSEN, R., and WESTLING, H.: On the accuracy of indirect auscultatory blood pressure measurements during exercise. Acta med. scand., Suppl., 449, 81 (1966)
KRÖNIG, B., DUFEY, K., REINHARDT, P., WITZEL, U., GRAULICH, M. und JAHNECKE, J.: Telemetrische Langzeitmessungen zur pharmakologischen Beeinflußbarkeit des Belastungsblutdruckes Hochdruckkranker. In: DISTLER, A. und WOLFF, H.P.: Hypertension, Symposium Mainz 1973; Thieme, Stuttgart 1974
KRÖNIG, B., MOERGEL, K., JACOB, H., GRAULICH, M. und JAHNECKE, J.: Blutdrucktelemetrische Untersuchungen zum Vergleich von Basis- und Entspannungsblutdruck bei Hochdruckkranken. Verh.dtsch.Ges.inn.Med. 79, 781 (1973)
KRÖNIG, B., PARADE, D., SCHWARZ, W., WITZEL, U., KLEMEIT, R., JAHNECKE, J. und WOLFF, H.P.: Blutdrucktelemetrie beim Menschen mit der Mikrokathetermethode. Klin.Wschr. 50, 898 (1972)
LITTLER, W.A., HONOUR, A.J., SLEIGHT, P., and STOTT, F.D.: Continuous recording of direct arterial pressure and electrocardiogram in unrestricted man. Brit.med.J. 3, 76 (1972)

Author's address: Ass.-Prof.Dr.med.B.Krönig, I.Medizinische Klinik und Poliklinik, Johannes-Gutenberg-Universität, D-65 Mainz 1, Langenbeckstr. 1, Germany

Biotelemetry II. 2nd Int. Symp., Davos 1974, pp. 216–218 (Karger, Basel 1974)

Multi-Channel EEG Telemetry-Computer Monitoring of Epileptic Patients

J.R. Ives, C.J. Thompson, P. Gloor, A. Olivier and J.F. Woods
Department of Neurology and Neurosurgery, Montreal Neurological Institute,
McGill University, Montreal, Quebec

The practice of recording and monitoring the EEGs of epileptic patients with bio-telemetry systems has been developing at the Montreal Neurological Institute over the last 3 years.

In our early use of telemetry a 4 channel radio-telemetry system was employed and is still used in research studied involving petit-mal 'absence' patients with generalized spike and wave discharges in their EEGs (Ives, Thompson and Woods, 1973).

Its usefulness in recording focal seizures was questionable, but with expansion to 8 channels more coverage of each hemisphere, particularly the temporal lobes,was available and the recording of clinical seizures was more informative. (Woods, Ives and Robb, 1973).

With the beginning of an implanted electrode program to determine more precisely the site of origin of intractable seizures in patients who were previously diagnosed as bitemporal lobe epileptics the need of even more channels was essential.

After several seizures (400) had been recorded in the first 4 patients with implanted electrodes while using the 8 channel radio-telemetry system, the following observations were made:

1) seizures from the same patient had strikingly similar on-set patterns in terms of frequency and channel of on-set.

2) the recordings were remarkably clear and free of movement, muscle and eye motion artifact even during a convulsive seizure.

These characteristics prompted us to investigate automatic computer monitoring and detection techniques which would eliminate the need for the technician, and the continuous paper record, or the necessity of an expensive and complicated tape recorder/playback system.

Our PDP-12 computer was programmed to monitor the 8 channels of subcortical EEG as received from the radio-telemetry system and used in this role whenever it was free during nights and weekends. The detection of the seizure was provided by monitoring the output of a bandpass filter (set to the dominant ictal frequency for that particular patient).

The delay of the EEG in the computer's memory at this time was only 5 seconds but the system was successful in the next two patients because it automatically recorded over 30 seizures which mainly occurred at night or on the weekend.

Because of the limited coverage of 8 channels when as many as 52 recording sites were available, an amplitude time division multiplexing system, similar to that outlined by Olsen, et al (1971) and Vreeland and Yeager (1973),was developed to transmit 16 channels of EEG via a coaxial cable from the patient's head to anywhere in the hospital. This allowed simultaneous coverage of 16 channels while the patient was only restricted by a 30' coaxial cable (radio-telemetry is being developed).

At the same time the computer's disc was used as temporary storage for the EEG signals enabling the 16 channels of EEG to be delayed up to 4 minutes (at 100 samples per second per channel). This allowed a seizure push button to be located on the patient's bed that could be pushed when a clinical seizure was obvious to the nursing staff or the patient felt a seizure beginning (aura).

The success of the implanted electrode monitoring system promoted its use on standard surface EEGs for the monitoring of seizure patients when it was desirable to record an ictal event for possible EEG localizing information.

Again the seizure push button, cable-telemetry and computer system were highly successful in recording the elusive ictal events with the following benefits:

1) the lack of movement artifact provided a much more informative EEG record.

2) several seizures were recorded during sleep and their initial on-set seemed to have significantly less muscle artifact than seizures recorded when the patient was awake. These features can be seen in Figure I.

Up to now three patients with implanted wires have been monitored with the 16 channel cable-telemetry-computer system for a total of 25 seizures, while the surface system has been used on 10 patients and over 25 seizures have been recorded.

Several important properties of the system are as follows:

Number of channels: 16
Weight of unit on head: 80 gms
Size: 10 X 5 X 3 cm.
Bandwidth: 0.5 Hz - 35 Hz
Noise level: 2 - 3 μVpp

Sample rate: 200 samples/sec/chan.
Input impedance: >1M
Common mode rejection: >100db
Battery life: 3 weeks

References:

IVES, J.R., THOMPSON, C.J. and WOODS, J.F.: Acquisition by telemetry and computer analysis of 4-channel long-term EEG recordings from patients subject to "Petit-Mal" absence attacks. Electroenceph. clin. Neurophysiol., 34: p. 665-668 (1973)

OLSEN, D.E., FIRSTENBERG, A., HUSTON, S.W. DUTCHER, L.R. and ADEY, W.R.: National Telemetry Conference, p. 308-312 (Washington, 1971)

VREELAND, R.W. and YEAGER, C.L.: A nine channel "COS/MOS" EEG telemetry for long-term seizure monitoring. 26th Ann. Cong. Eng. in Med. & Biol., p. 402 (Minneapolis, 1973)

WOODS, J.F., IVES, J.R. and ROBB, J.P.: The application of telemetric recording of the electroencephalogram to the clinical management of epilepsy. VIIIth Inter. Cong. of Electroenceph. clin. Neurophysiol. p. 709 (Marseilles, 1973).

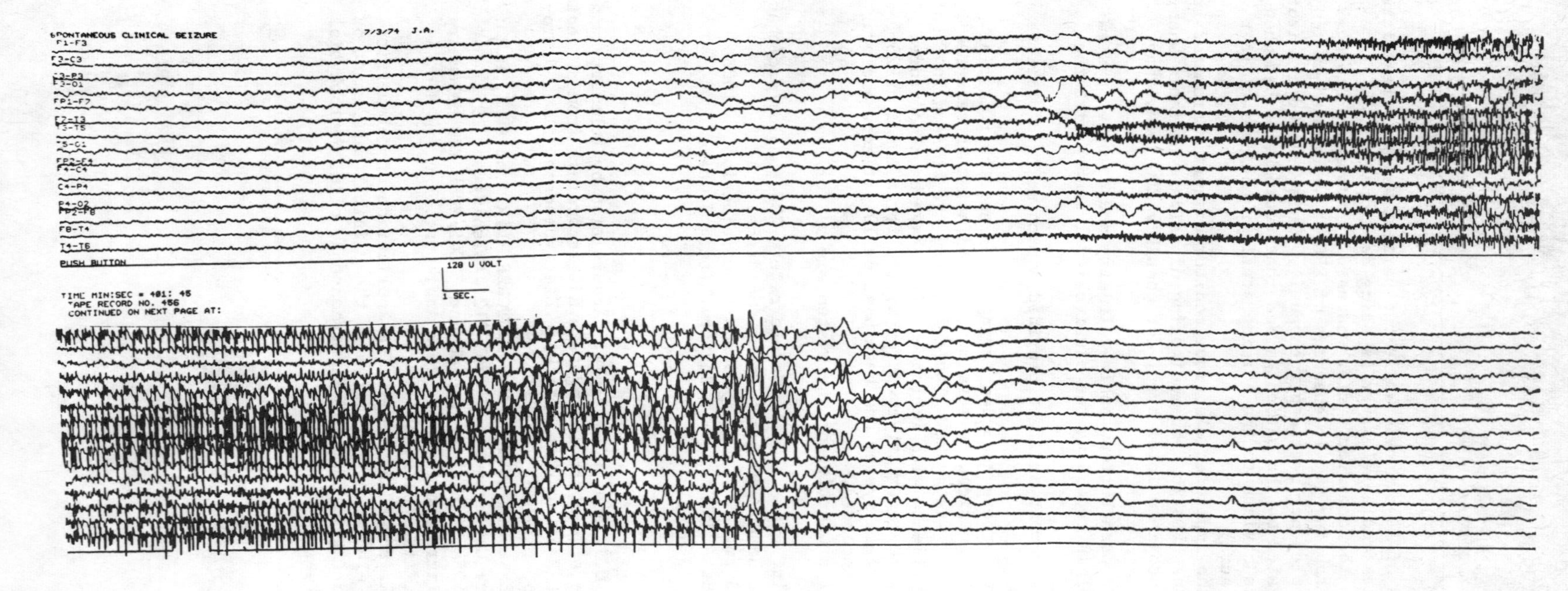

Figure I: The above illustrates the second of five surface seizures as photographed from the computer screen (composed of 6 separate 10 sec. epochs) recorded on a 39 year old male patient. This seizure occurred at 11:09 p.m. on March 6/74 while the patient was in a drug induced sleep state. His obvious clinical manifestations alerted the nurse who then pushed the 'seizure push button' located near the patient (this action can be seen on the last channel, lower left hand corner); this in turn activated the computer program to permanently store 4 minutes of EEG prior to the pushing of the button and 1 minute after. The elapsed time(401:45) in minutes and seconds is also recorded to allow correlation with the actual clinical event recorded in the nurse's notes.

The recording is from silver/silver-chloride scalp electrodes (channel 1-8 left hemisphere; 9-15 right hemisphere; 16 'seizure push button').

Session IX

Long-Distance Telemetry and Tracking

Chairman: *H. U. Debrunner*

Biotelemetry II. 2nd Int. Symp., Davos 1974, pp. 220–222 (Karger, Basel 1974)

Microwave Techniques in Animal Radiotelemetry

William Keith Brockelsby
School of Physics, University of Sydney, Sydney, NSW

In mid 1971 an Environmental Physics department was established within the School of Physics of the University of Sydney. The Radiotelemetry Group, the first within the Department, was set up to carry out development of equipment for use in ecological and biological studies of free ranging wildlife species.

A full scale long term study of Australias' saltwater crocodile *Crocodilus porosus*, was just beginning, and the first task assigned to the Group was that of producing a radio tracking system capable of monitoring the movements of individual crocodiles, possibly over very considerable ranges and for periods of several years.

By 1971 several projects based upon the use of long range tracking and telemetry had been undertaken by other investigators.

In space research telemetry techniques had reached an advanced stage of reliability, but in terrestrial tracking and biotelemetry the record was not nearly so brilliant.

It is not the purpose of this report to argue in detail the merits of the various options and limitations facing designers of such equipment, but simply to emphasise the principles underlying my approach to the problem and to describe the results of this work.

SYSTEM DESIGN CONSIDERATIONS.

Almost all biotelemetry transmitters and tracking beacons must be small and unobtrusive, and in general the designs have incorporated efficient low power electronic circuits limited in life by the power source. These small devices each include an "antenna", or radiating element which is also very small, and when its size is considered in relation to the wavelength of the energy normally used, it is found to be extremely inefficient.

Most of the energy is dissipated around, and not radiated from, these small, low impedance antennas. If longer wires or whips are used to increase the length, the risk of damage is greatly increased. A number of attempts at producing fractional wavelength resonant structures have been made with some success, but these are likely to become severely detuned by changes in environmental conditions and by the proximity of nearby objects. At the outset of our programme we sought to avoid these problems by investigating the use of very much shorter wavelength radiation. Even for an animal as large as a crocodile, a tracking transmitter should not exceed approximately 6cm in width and the profile should be as low as possible. We decided on a height limit of 4cm, of which 3cm could be allowed for the antenna.

This meant that the operating frequency should not be less than 1200MHz for efficient radiation.

The next step was to determine whether a system using this frequency, (wavelength 25cm):

(a) could be designed with adequate communications performance under all required conditions.

(b) could be built within the size and weight constraints imposed.

The first question was answered theoretically to a first approximation by assuming the following nominal design values,

<u>Receiver</u> Noise Figure 3.5db, Bandwidth 3kHz, Sensitivity - 135dbm. Receive antenna gain 10db

<u>Transmitter</u> Output Power 10mW, Frequency Stability within 6kHz. Transmit antenna gain Odb.

The overall system could thus accommodate a transmission loss of 155db. It was assumed that aircraft would be required to successfully carry out wide-ranging observations on populations in inaccessible areas, so that long range calculations could be made assuming the receiver is at 3,000 metres altitude. Assuming also that the effective height of the transmitting antenna in saltwater surroundings would be about 30cm, then a range of more than 200 Km should be attainable. The effect of terrain, objects and surface roughness in the immediate vicinity of the transmitter would of course considerably modify the radiation pattern and reduce the maximum range.

The second question was not so readily answered, but after building a variety of circuits in different configurations, satisfactory models for both transmitter and receiver were finally developed. The use of microstrip techniques was the key to success here.

With basic transmitter and receiver circuits available, a wide range of uses (including tracking and data transmission) becomes possible with appropriate additional circuitry.

For a tracking system, a coded pulse generator was required to provide for identification of individual transmitters and to control the rate of usage of power. If the pulse rate could be controlled, additional facilities, such as regulation of the output power in proportion to the energy available from a solar cell, or transmission of the temperature of a battery for instance, would be available. CMOS integrated circuits, with their negligible power requirements, were to be used where-ever possible. Two forms of power sources were to be produced, one based on Lithium primary cells yielding a finite life of, say, 6 or 12 months, and one using silicon photovoltaic cell arrays and a Ni-Cd rechargeable battery. This solar powered model was to operate for an indefinite period, hopefully for several years.

<u>SYSTEM PERFORMANCE AND FIELD RESULTS</u>.

The performance assumed for both transmitter and receiver in their design stage was met in actual practice. The transmitter antenna finally adopted was a vertically polarised folded monopole and small ground plane. The monopole element was effectively reduced in height by enclosing it in teflon and loading it with a capacitive disc. The resulting impedance was near 50 ohms resistive and the radiation pattern donut-shaped with a central vertical null, very useful in overhead tracking operations. The overall transmitter efficiency from DC to RF was 10%, the antenna efficiency being about 90%. The transmitter-antenna assembly measures 6.2cm diameter by 4.0cm height and weighs 25gm. A typical crocodile transmitter powered by Li-cells for 6 months, weighs 230gm, of which 30gm is Li-cell, 35gm is electronics and 165 gm is packaging.

The system has been extensively tested in many ways and under difficult field conditions on both crocodiles and on dingo (Australian native wild dogs). Successful tracking of both species has been achieved, and aside from one early problem with a leaking package, the transmitters have been recovered later in good condition and fully operational. No degradation of performance has been observed under field conditions.

Considerable experience has been gained in operating with the system under a variety of conditions. Typical ranges obtained are 5Km over flat wooded ground, 10Km or more from vantage points in hilly country using hand-held receivers, and better than 200 Km over sea from an aircraft. Crocodiles in mangrove creeks have been accurately located from 30 Km from aircraft at 1000 metres. Even in an inner city area ranges of around 1 - 2 Km are readily attainable from ground tracking.

FURTHER DEVELOPMENTS.

Aside from a possible improvement in transmitter efficiency to about 30% using an alternative design, some further reduction in size may now be expected by using even higher frequencies. However very little significant improvement in transmitters is anticipated without incorporating hybrid microcircuit technology in which circuit dimensions at microwave frequencies become attractive. Our department is committed to investigating these proposals, and to producing its own microcircuit packages using thin film techniques.

It is now planned to produce a complete long-range system for transmission of physiological and other data, based upon a general purpose microwave system, and relying upon extensive use of microcircuit technology in order to miniaturize the volume of practical data-transmission packages.

Biotelemetry II. 2nd Int. Symp., Davos 1974, pp. 223–225 (Karger, Basel 1974)

Instrumentation for Studying Social Activity in Mouse Colonies

John A. Henry, Roland D. Rader, Daniel L. Ely and James P. Henry
University of Southern California, School of Medicine, Department of Physiology, Los Angeles, Calif.

To evaluate behavioral influences on cardiovascular functions, there is a need for a detailed and quantitative method by which the drives for food and water and sexual and territorial interactions can be continuously monitored in groups of animals living in population cages. This method can help to establish the relation between behavioral phenomenon and the progression of disease; it can also be of value in the study of behavioral pharmacology.

With this goal in mind, we have developed a technique in our laboratory that relies on the detection of a magnetic field generated by 9 x 2 mm cylindrical magnets implanted in male and female mice, socially interacting in colonies. The mice are placed in an "open field" container 270 x 270 x 13 cm. Two 23 x 11 x 11 cm standard polycarbonate mouse cages are attached to the sides of the open field by 3.8 cm acrylic tubing and each connecting tube is fitted with a special detection portal. The small side-cages are differentiated into functional areas for food, water, nesting, and activity.

The tiny Alnico VIII magnets are implanted in the back or in the abdomen of the mice or in both sites simultaneously. When the possibilities of these two locations are combined with the capability of polarizing and depolarizing each magnet at will, in situ, the coding of eight individual mice becomes feasible.

The magnetic detection circuitry uses Hall Effect devices to detect the presence, position, and polarity of the magnets during the passage of a mouse through a portal in either direction. These circuits consist of operational amplifiers and comparators that condition the signal of the Hall devices to standard TTL 5-volt logic levels. This logic fills a binary register which displays the unique coding for a mouse making the passage.

In an earlier paper (1), a manual system of clocks and counters was used to follow the activity of only two mice. The ability to code and follow eight animals simultaneously and continuously quadruples the data flow, making it necessary to automate the data presentation and analysis.

A CIP 2100 minicomputer logs the time and location of each mouse's activity on punched paper tape. The paper tape data is processed by a Fortran IV program on an IBM 370/155, and the passage of mice into functional areas--food, water, etc.--is converted to the percent of elapsed time in each area. The number of transactions made between the areas is also counted. This data is presented in the form of 1-hour, 6-hour, and 24-hour summaries. The 6-hour summary is then used to make a computer plot of a graphic representation or behavioral map (Fig. 1) of each animal's activity in the colony during the desired time interval. The area of the circle represents the percent of time the mouse spent in that location and the width of the paths between the functional areas is proportional to the number of times the mouse made such a trip. These behavioral maps or ethograms provide quantitative and continuous measurements of an animal's activity patterns.

In order to evaluate the behavioral monitoring system sensitivity, several drugs were tested on high activity male mice that were socially interacting in a population cage. Figure 1 shows the behavioral effects of a new beta blocker (H 73/26 - Hässle Pharmaceuticals, Mölndal, Sweden). The drug did not cause a major change in the amount of time spent feeding; however, there was a slight reduction of time in the water area and no time was spent with the females. Also locomotor activity decreased from 19 to 6 entry-exits per hour during the 6-hours of posttreatment. During the following 6 hours, the behavior patterns and locomotor activity returned to control values. Other drugs, such as marihuana, amphetamine, and cyproterone acetate, have been behaviorally tested in our laboratory and they have all produced changes in rodent behavior patterns as measured by this technique.

This behavioral monitoring system offers great potentiality in that it not only detects changes in locomotor activity, as in photocell systems, but it also detects changes in individual patterns of behavior that show the same locomotor activity. This technique is presently being used in drug environment, brain lesion, and ethological investigations.

Reference

ELY, D. L.; HENRY, J. A.; HENRY, J. P.; and RADER, R. D.: A monitoring technique providing quantitative rodent behavior analysis. Physiol. Behav. 9: 675-679 (1972).

This research was supported by the National Institutes of Health Grant No. MH19441.

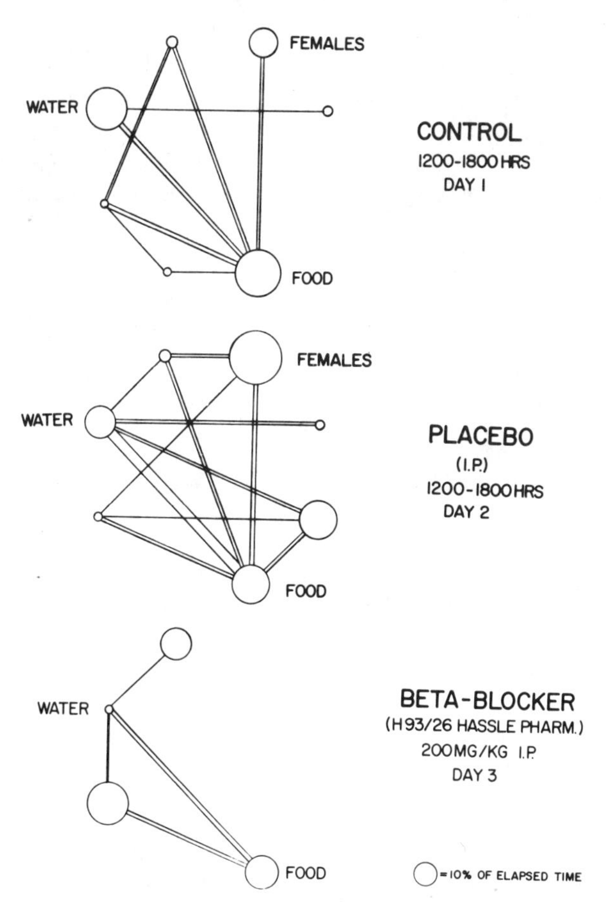

Figure 1. Behavior patterns of a high activity male in an 8-box population cage during a control period, after a placebo injection of saline and after administration of the beta blocker. All behavioral measurements were made during the hours 1200-1800. The area of each circle is proportional to the time spent there and the width of the path is proportional to the number of transactions made between the two boxes.

Biotelemetry II. 2nd Int. Symp., Davos 1974, pp. 226–228 (Karger, Basel 1974)

Telemetry of Electrocardiograms from Free-Living Birds: A Method of Electrode Placement

James A. Gessaman

Biology Department, Utah State University, Logan, Utah

In 1970 I began an evaluation of heart rate as an indirect measure of the energy metabolism of free-living birds. The ECG of a bird was telemetered to a remote site and recorded on a strip chart recorder. Minute heart rates were determined by counting the number of QRS complexes occurring during a one-minute ECG recording. Monitoring heart rate from a bird in the wild requires an electrode placement that is not easily accessible to the bird's mandibles and feet and that can remain functional for several weeks to months. In addition, electrode placement must give a discernible ECG signal while the bird is at rest as well as during flight when strong EMG potentials from the active pectoral muscles are present. I experimented with several electrode placements in the American kestrel (Falco sparverius) and found the following one to be superior.

Two electrodes are anchored into the keel of the sternum near the heart from inside the body cavity. The electrode leads are brought through an incision in the abdominal muscle wall and threaded under the skin from the abdominal incision to the bird's back. They then pass through the skin and terminate at a position that is not readily accessible to the bird.

This procedure will now be described in detail. An incision about 6 mm long is made just posterior to the sternum through the midline of the abdominal wall of an anesthetized bird. An electrode, consisting of a No-Knot Eyelet (available at sporting goods stores) silver soldered to 8 cm of very flexible, multistrand, teflon-coated, stainless steel wire (available from Narco Bio-Systems, Inc., Houston, Texas), is grasped firmly with a hemostat and inserted through the incision. It is carefully slipped along the dorsal surface of the sternum (parallel to the keel) to a position near the apex of the heart. (It is helpful to practice on a dead bird before attempting a live implant). The point of the No-Knot Eyelet is then thrust into the keel while the breast of the bird is braced anteriorly with the other hand. The second electrode is anchored similarly in the midline of the sternum 1 to 2 cm posterior to the first (fig. 1A & B). The muscle of the abdominal wall is then stitched closed around the electrodes with triple zero chromic sutures; one or two stitches are usually sufficient.

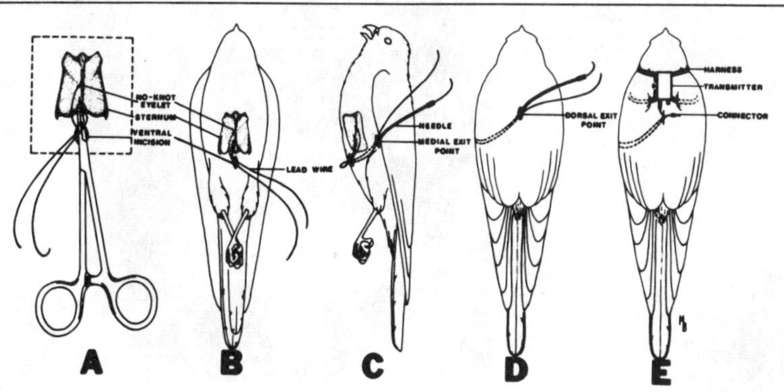

Figure 1. A and B, a small incision is made through the abdominal wall and the two electrodes (No-Knot Eyelets) are pushed into the keel of the sternum. C, a curved 15-gauge hypodermic needle is inserted under the skin medially and anterior to the femur and guided through the subcutaneous tissue to the abdominal incision. Both lead wires are threaded through the needle, then the needle withdrawn. D, the step illustrated in 1C is repeated, threading the wires from their exit point in C to the middle of the back. E, the harness and transmitter are shown in place. The exit point of the wires should be under the transmitter for maximum protection from the bird's mandibles.

The two electrode leads are then threaded under the skin to the bird's back through a curved 15-gauge hypodermic needle. More specifically, the needle is inserted under the skin on the bird's side, just anterior to the femur, and guided through the subcutaneous tissue to the incision. The wires are then threaded into the tip and through the bore of the needle. The needle is withdrawn leaving the wires under the skin (fig. 1C). The abdominal incision in the skin can now be sutured closed. This same procedure is repeated, starting from a point in the middle of the bird's back and going to the lead wires which protrude from the skin on the side (fig. 1D).

Connectors compatible with those on the transmitter are soldered to the lead wires protruding through the skin on the back. A harness was used to secure the ECG transmitter to the bird (fig. 1E & 2). The lead wires were brought through a slit in the leather saddle of the harness. When the transmitter was not in place on the harness, the exposed electrode leads were taped to the harness to protect them from the bird's mandibles.

Figures 3A & B are X-rays of a kestrel with electrodes and harness (without a transmitter) in place.

This method of electrode placement has several features which are superior to fastening electrodes onto the skin of the bird's dorsal surface or of its pectoral region with suture thread or a safety pin. One, the electrode leads are subcutaneous and inaccessible to the bird's mandibles or feet except for a short piece (about 2 cm) that extends through the skin and connects to the dorsally-mounted

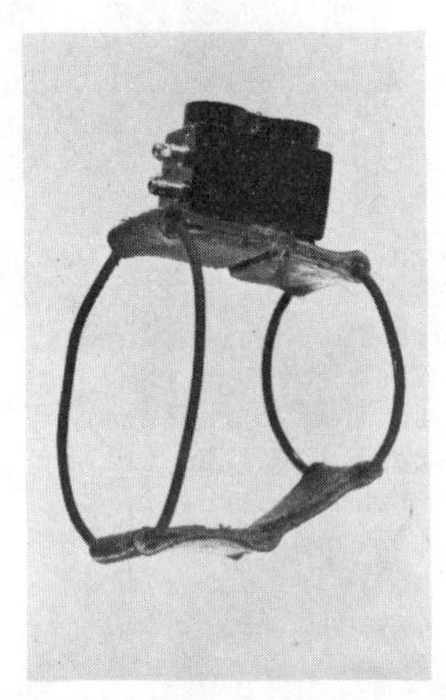

Figure 2. An ECG transmitter mounted on a harness fitted for a kestrel.

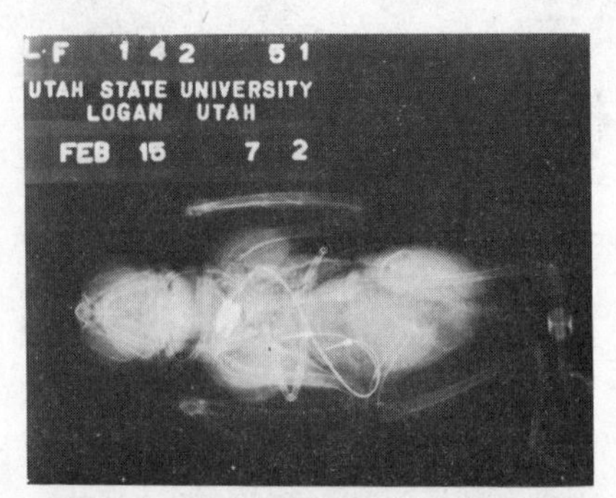

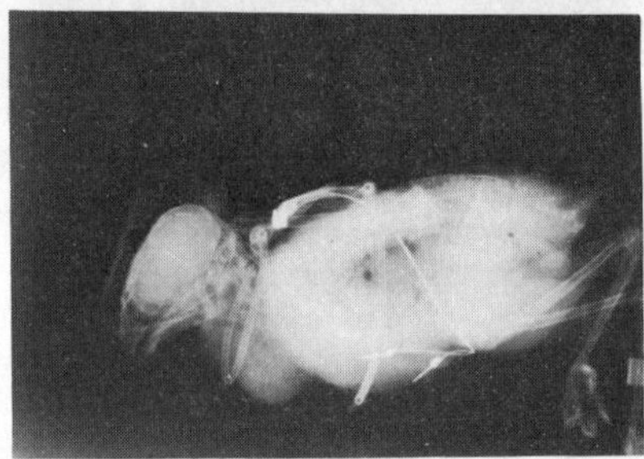

Figure 3. An X ray of a kestrel with two electrodes in its keel and lead wires extending to connectors on its dorsal surface. The bird is wearing a harness. A, the bird's dorsal surface is facing the radiation source. B, the bird's left side is facing the radiation source.

transmitter. Two, electrocardiograms recorded from the keel electrodes during flight have much less EMG interference than those recorded by surface electrodes. The reduced interference may be due to multiple factors. Among these may be that the sternum bone shields the keel electrodes from the EMG potentials better than subcutaneous tissue shields surface electrodes and that the EMG potentials along the keel are determined by the electrical activity in both pectoral muscles, the EMG's from the pectorals may tend to cancel out along the keel which separates these two muscles. Three, the electrodes are anchored into the bird's skeleton, thus rigidly fixing the distance between them even though the bird is in flight. Four, the electrodes will remain in the same position for months in a free-living bird.

With this method we have telemetered ECG's from birds in the field for two-to three-week periods. One kestrel was implanted and released into the field. Nine months later it was retrapped and the electrodes were still functional.

This work was supported by NSF Grant GB-18158.

Biotelemetry II. 2nd Int. Symp., Davos 1974, pp. 229–232 (Karger, Basel 1974)

A Long Range Emergency Telemetry Link

J.R. Jonkers and J.G.M. Kersemakers
St. Annadal Ziekenhuis, Afd. Funktielaboratorium, Maastricht

Introduction: In the Netherlands, the law on paramedics prohibits the independent use of defibrillators by nurses and ambulance crews. These therapeutics are strictly reserved to medical doctors. However, a cardiologist can delegate his responsibility to the nurse, allowing her to do the defibrillation. He doesn't have to do it with his own hands! After studying the literature on this subject, the authors found out that radio telemetry for the E.C.G. and radio telephonic contact was enough cover for the ambulance attendants. The decision to use the Ambulance Radio Telephone (A.R.T.) as E.C.G. transmission medium prompted the authors to search for the most desirable solution. These A.R.T.'s operate on one frequency (Simplex) and after several experiments we found out that this was not the ideal type of communication for this purpose. Especially the "over procedure" commonly used in simplex traffic was difficult to maintain in stress situations, making it otherwise necessary to transmit E.C.G. and speech alternatively, with the possibility of information being lost; a system of duplex communication became necessary. Duplex is a radio communication system in which both transmitters transmit on two different frequencies, so that the method of communication is similar to normal telephone calls (no "over procedure"!).

System description: (Figure 1.) On the highest roof of the hospital (+40 m) an antenna 12 m high was built for transmitting-receiving purposes. In a duplex system, it is enough to use one antenna for both purposes by means of a duplex filter (in Fig. 1 D.F.), so that transmitter and receiver can operate simultaneously. At this relatively high altitude, a small powered (3 Watt) transmitter is sufficient to reach the ambulance in the field. Conversely, for the ambulance returning home, perhaps because of the absorbent environment of city buildings, it is necessary that the ambulance transmitter be more powerful in order to reach the receiving antenna, and is therefore rated at 10 Watts. Let us now consider the situation in which the ambulance crew has to operate, for example, at the heart patient's home. First they leave the car and bring the stretcher to the patient, making him ready for dispatch. Many times in the past they found a patient with evident circulation problems and they could not give adequate help, which is nowadays possible by means of cardiac monitoring and defibrillators. The moment they left the ambulance, communication with the hospital was lost; it was therefore necessary to retrieve communication in the distance between the car and the sickbed of the patient. For this purpose the ambulance personnel is now equipped with a portable transceiver of 500mW transmitting power. The A.R.T. in this circumstance acts as a relay station (switch Fig. 1), with the same frequency as the main communication system. The E.C.G. is transported to the car by a pocket-size 430 Mhz, 2 m Watt commercially available transmitter. In the car, the 430 Mhz carrier is detected by a suitable receiver, and after demodulation visible on an E.C.G. monitor. This situation will be maintained until the patient is in the C.C.U. in the hospital. After demodulation of 430 Mhz carrier, the E.C.G. is modulated again on a specially designed modulator. Then it is mixed with the signal in the voice channel of the main duplex transmitter. At this moment, voice and E.C.G. travel simultaneously to the hospital. In the receiver at the hospital, this signal is split out in E.C.G. and speech again. Eventually from here all the information can be brought to the physician at home by means of a normal telephone. When the patient is placed upon the stretcher and equipped with the pocket-size E.C.G. transmitter, the patient is ready to dispatch. In this way the ambulance crew have their hands free to maneuver the stretcher and are still in contact with the hospital by means of the small transceiver, carried by one man. In the ambulance, the communication described above is maintained, with the exception that the main A.R.T. no longer acts as a relay station but operates normally (switch S Fig. 1).

<u>Technical description of modifications in the transmitter:</u> (see Fig. 2) The authors studied the overall frequency content of human voices over the A.R.T. system by a Fourier-analysor, and discovered that the realistic speech-spectrum (200-2000Hz) of several human voices almost never exceeds 2000Hz. Furthermore, they found out that by direct modulating (also eliminating all amplifiers) of the transmitter, a transmission of 8000Hz. was possible. There is also a small surplus bandwidth available from 2000 to 8000 Hz. In this area it is possible to do transmissions of E.C.G.'s by a suitable sub-carrier. Of course, in practice the 8000 Hz limit may never be approached, to avoid cross-talk in the neighboring channels of other A.R.T. users by extreme side-bands development. A safe limit grounded on P.T.T. (Post Telegraph Telephone Company) rules on this theme is approx. 4000 Hz, so that the real bandwidth is approx. 3000 to 4000 Hz. In this case the authors chose a center-frequency of 3600 Hz for the narrow-band F.M. sub-modulator. Also, it was necessary to prevent the main modulator from too deep modulation by the amplitude of sub-carrier. This was realized by a solid state switch between sub-modulator and main-modulator which isolates above the predictable level of 150 m.V.p.p.. To improve the separation in the receiving station between the speech and sub-carrier, it is necessary to filter (cutoff above 2000 Hz) the speechchannel in the transmitter side.

Technical description of the receiving station (see Fig. 3). In the receiving station at the C.C.U., it is relatively simple to split speech and E.C.G. because the transmitter stage already takes care that the spectra of voice and E.C.G. subcarrier are far away from each other. In the detection of the E.C.G., speech is considered as noise on the subcarrier. Only zero-crossing is important for the subcarrier. The discrimination between noise and subcarrier is easily obtained by a phase locked loop (P.L.L.) with a lock range of 3500-3700 Hz. The P.L.L. is a part of the versatile integrated tone decoder of Exar type XR- 567, which can easily be programmed in frequency and bandwidth by simple external components. Furthermore, it produces a logic output when the P.L.L. is locked on a signal within the predicted frequency range. The authors used this possibility as a very sharp switch (squelch) which commands the visibility of E.C.G. on the monitor. In practice, it can happen in some cases that the transmitting antenna of the ambulance is no longer "seen" in the receiving station, and the built-in squelch of the A.R.T. is not yet active. In these cases, it is absolutely necessary to prevent the subcarrier from being fed with too-noisy signals. To explain this, one must first consider the demodulator type used. The output of the P.L.L. is connected to a limitter followed by an overdriven amplifier to shape the rise time of the created block wave. After integrating, the original E.C.G. is available as an amplitude modulated signal, with a triangle carrier. This carrier is simply eliminated by another Butterworth 4 pole active filter, which cuts off above approx. 50 Hz. By this technique and with a high noise

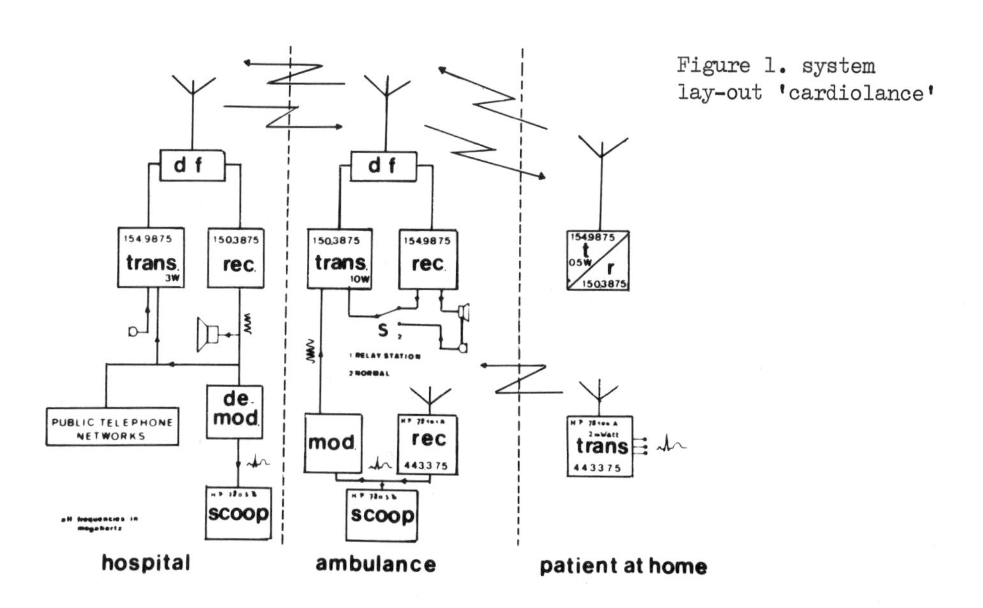

Figure 1. system lay-out 'cardiolance'

content in the subcarrier (the XR 567 is now fortunately unlogic!) the demodulated E.C.G. looks very similar to ventricular fibrillation, so that the unlogic output of the P.L.L unblanks the E.C.G. on the monitor and an acoustic signal announces: "the transmitter is out of range." In the speech channel everything is simple, the sub-carrier of 3600 Hz is eliminated by a low pass filter, but stays just audible to the operator as an extra guide against arithmias.

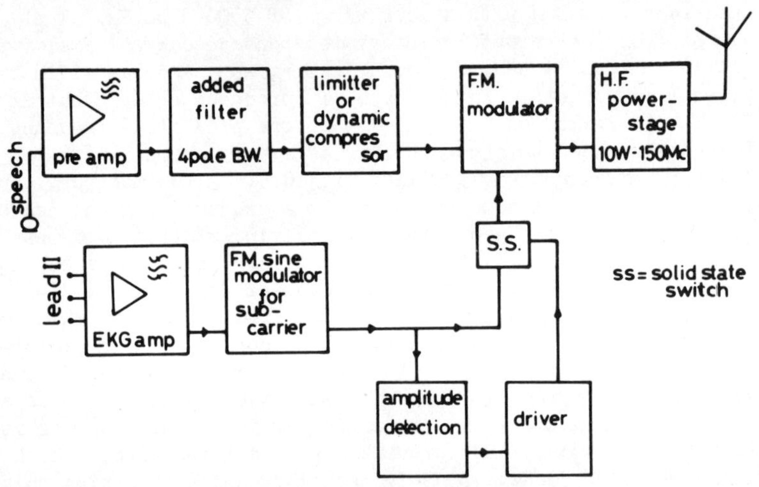

Figure 2. Modification duplex mobilophone transmitter for speech and E.C.G. transmission simultaneously.

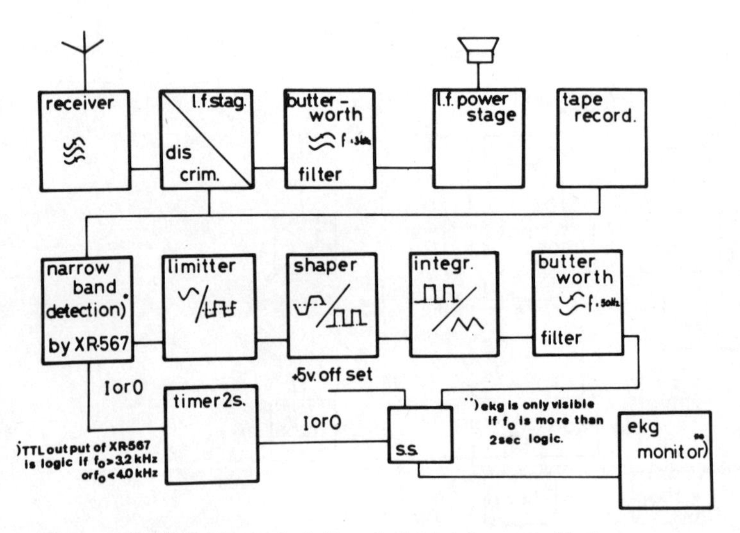

Figure 3. Receiver C.C.U. station for cardiolances.

Appendix

List of Participants

Frau Ch. Alexandroff
Medical Tribune
Burgstr. 6
D–6200 Wiesbaden (FRG)

Dr. E. Asang
TU München
Klinikum rechts der Isar
Belgradstr. 5
D–8000 München (FRG)

Dr. K. Bachmann
University of Erlangen
Medizinische Universitätsklinik
D–8520 Erlangen (FRG)

Mr. L. Bakema
Electr. Eng. Res. Dept.
St. Antonius Hospital
Jan v. Scorelstraat 2
Utrecht (The Netherlands)

Mr. J.O. Bakker, El. Eng.
University of Groningen
Burg. Ritzemastr. 44
Niekerk (The Netherlands)

Mr. A. Balhy
Institut des métaux et
des machines
Av. de Bellerive 34
CH–1000 Lausanne (Switzerland)

Dr. W. Bauer
Medizinische Biologie
Sandoz AG
CH–4002 Basel (Switzerland)

Prof. Dr. E. Baumann
Swiss Federal Institute of
Technology
Institute of Applied Physics
Postfach
CH–8049 Zürich (Switzerland)

Dr. J.U. Baumann
Department of Orthopedic Surgery
Kinderspital
Römergasse 8
CH–4005 Basel (Switzerland)

Mr. R. Baumgartner
Kinderspital
Römergasse 8
CH–4005 Basel (Switzerland)

Dr. P. Baurschmidt
Abteilung für Biomedizinische
Technik
Friedrich-Alexander-Universität
Turnstr. 5
D–8520 Erlangen (FRG)

Dr. W. Becker, Eng.
CCR Euratom
I–21027 Ispra (Italy)

Dr. J. Bleeker
Nipg TNO
Wassenaarseweg 56
Leiden (The Netherlands)

Dr. N. von Blumenthal
Biotechnik-Laboratorium
Chirurgische Universitätsklinik,
West Freie Universität Berlin
Spandauer Damm 130
D–1000 Berlin 19 (FRG)

Dr. B. Bohus
Rudolf Magnus Institute for
Pharmacology
Medical Faculty
University of Utrecht
Utrecht (The Netherlands)

Dr. J. Bojsen, B.Sc. El. Eng.
The Finsen Institute
Strandboulevarden 49
DK–2100 Copenhagen (Denmark)

Mr. K. Boitchev
Scientific Sport Centre CNIFK
bul. Tolbuhin 18
Sofia (Bulgaria)

Mr. G. Bomben
Forschungsdirektion
Psychiatrische Universitätsklinik
CH–8029 Zürich (Switzerland)

PD Dr. A.A. Borbély
Institute of Pharmacology
Gloriastr. 32a
CH–8006 Zürich (Switzerland)

Mr. K.H.N. Bos, Eng. Researcher
University of Groningen
't Zuden 79
Leek (The Netherlands)

Mr. W.C. Bosshardt
Ing. Tech. HTL
Abteilung für Experimentelle
Orthopädie
University of Bern
Inselspital
CH–3010 Bern (Switzerland)

Dr. J. Boter
Institute of Medical Physics TNO
45, Da Costakade
Utrecht (The Netherlands)

Dr. J. Brennwald
Forschungsinstitut
CH–7270 Davos (Switzerland)

Mr. K. Bretz
Research Institute of Physical
Education TFKI
Alkotas u.44
H–1123 Budapest (Hungary)

Mr. W.K. Brockelsby, Proj. Eng.
The University of Sydney
School of Physics
Sydney, NSW 2006 (Australia)

Mr. R. Bühler
Leiter der Abteilung
Instrumenten-Technik
Byk-Gulden Pharmazeutika
Byk-Gulden-Str. 2
D–7750 Konstanz (FRG)

Mr. D.A. Burchard, Dipl. Ing.
Ingenieurbüro Burchard
Tiroler Weg 2
D–7801 Ebringen (FRG)

Mr. H.H. Büscher
Sandoz AG
Gebäude 881/311
CH–4002 Basel (Switzerland)

Dr. R. Canivec
Directeur de CEBAS-Pr., CNRS
Villiers-en-Bois
F–79360 Beauvoir (France)

Prof. V. Cappellini
Instituto di Ricerca sulle Onde
Elettromagnetiche
Via Panciatichi 56
I–50127 Firenze (Italy)

Dr. S. Carbonini
Medi Metron S.p.A.
Via Rutilia 23
I–20141 Milano (Italy)

Mr. D.J. Cathignol
INSERM
51, Av. Condorcet
F–69100 Villeurbanne (France)

Mr. J. Cordey
Schweizerisches Forschungsinstitut
CH–7270 Davos (Switzerland)

Mr. Covegnas
SFENA
F–78140 Velizy (France)

Mr. D.E. Creighton
British-American Tobacco Co. Ltd.
Group R+D Centre
Regent's Park Road
Southampton SO9 1PE (England)

Dr. H. Czerny
Universitäts-Frauenklinik
D–4000 Düsseldorf (FRG)

Dr. H.U. Debrunner
University of Bern
Department of Experimental
Orthopedics
CH–3010 Bern (Switzerland)

Dr. Ing. F. del Polo
Departamento Fisiologia
Fac. Medicina
Universidad Autónoma
Herederos de Navas s/n
Madrid (Spain)

Mr. E. Denzler
Neurochirurgische
Universitätsklinik
Kantonsspital Zürich
Rämistr. 100
CH–8006 Zürich (Switzerland)

Mr. A.V. Deriabin
Kislovodsk Territorial Council
on Management by Health Resort
50 Years of October, 9
Kislovodsk 357700 (USSR)

Dr. R. Deroanne
Institut E. Malvoz
4, Quai de Barbou
B–4000 Liège (Belgium)

Dr. C.M. Dowse
Department of Oral Biology
University of Manitoba
Winnipeg R3E OW3 (Canada)
at present Pathologisches
Institut der Universität Bern
CH–3000 Bern (Switzerland)

Dr. G.U. Dumermuth
Kinderspital Zürich
Steinwiesstr. 75
CH–8032 Zürich (Switzerland)

Dr. Farago, Oberarzt
Schweizerische Anstalt für
Epileptische
Bleulerstr. 60
CH–8008 Zürich (Switzerland)

Dr. H. Fischler
The Weizmann Institute of Science
Rehovot (Israel)

Dr. H. Fleischer
Medizinische Poliklinik der
Universität
Östliche Stadtmauerstr. 29
D–8520 Erlangen (FRG)

Mr. C. Forni
CNRS
31, chemin 5 Aiguier
F–13274 Marseille Cédex 2 (France)

Dr. G. Foroglou
PD Neurosurgery
Hôpital Cantonale
CH–1011 Lausanne (Switzerland)

Dr. K.E. Forward
Department of Electrical
Engineering
The University
Newcastle-upon-Tyne
Newcastle-upon-Tyne NE1 7RU
(England)

Mr. G.R. Francis
Beecham Research Laboratory
Brockham Park
Betchwork, Surrey RH3 7HJ
(England)

Dr. med. B. Fränkel
Sandoz AG
Toxikologie 881/209
CH–4002 Basel (Switzerland)

Dr. H.M. Fromherz
Neutechnikum
CH–9470 Buchs (Switzerland)

Dr. Th.B. Fryer
NASA Ames Research Center
Moffett Field, CA 94035 (USA)

Dr. T. Furukawa
Research Institute of Applied
Electronics
Hokkaido University
N–12, W–6
Sapporo (Japan)

Dr. J.M. Gaillard
Médecin-chef de la recherche
Clinique psychiatrique
universitaire de Bel-Air
CH–1225 Chêne-Bourg
(Switzerland)

Mr. G.H. Gautschi, Dipl. Ing. ETH
Kistler Instrumente AG
Postfach 304
CH–8408 Winterthur (Switzerland)

Mr. U. Gebauer
Pathologisches Institut der
Universität
Hügelweg 2
CH–3000 Bern (Switzerland)

Dr. M.F. Gerold
F. Hoffmann-LaRoche & Co. AG
Grenzacherstr. 124
CH–4002 Basel (Switzerland)

Dr. J.A. Gessaman
Utah State University
UMC 53
Logan, UT 84322 (USA)

Mr. R.W. Gill
Stanford Electronics Laboratory
McCullough, Rm 26
Stanford, CA 94305 (USA)

Mr. H.W. Glonner
Glonner Electronic
Hofmillerstr. 2
D–8000 München 60 (FRG)

Mr. D. Godin, Ing.
Clinique Neuro-Chirurgique
Hôpital de Bavière
Université de Liège
Boulevard de la Constitution 66
B–4000 Liège (Belgium)

Mr. R. Grandjean, Cand. El. Ing.
Zollikerstr. 116
CH–8702 Zollikon (Switzerland)

Dr. G.C. Grigg
Zoology Department
University of Sydney
Sydney, NSW 2600 (Australia)

Dr. C.-M. Grimm
Klinikum rechts der Isar
Ländstr. 5
D–8000 München 22 (FRG)

Mr. R.W. Haberl
Österreichisches Institut für
Sportmedizin
Possingergasse 2
A–1150 Wien (Austria)

Mr. V. Halmo, Dipl. Ing.
Ciba-Geigy AG
K. 125.12.16
CH–4002 Basel (Switzerland)

Dr. J. Hanley
Space Biology Laboratory
Brain Research Institute
UCLA Health Sciences Center
Los Angeles, CA 90024 (USA)

Dr. V. Hanusová
Institute for Hygiene and
Epidermology
Srbarova 48
Praha 10 (Czechoslovakia)

Mr. K. Hausamann, Stud. Ing.
Wander AG, Forschungsinstitut
Monbijoustr. 115
CH–3001 Bern (Switzerland)

Prof. M. Hebbelinck
Secretary General ICSPE
Hoger Inst. voor Lichamelijke
Vrije Universiteit Brussel
Hegerlaan 28
B–1050 Brussel (Belgium)

Dr. O.W. Hess
Department of Obstetrics and
Gynecology
Yale University, School of Medicine
789 Howard Ave.
New Haven, CT 06510 (USA)

Mr. R.C. Hill, Pharmacologist
Sandoz AG
Geb. 881/332
CH–4002 Basel (Switzerland)

Dr. H. Hoernicke
Universität Hohenheim
Abteilung für Zoophysiologie
D–7000 Stuttgart-Hohenheim (FRG)

Dr. med. H. Howald
Forschungsinstitut der
Eidgenössischen Turn- und
Sportschule
CH–2532 Magglingen (Switzerland)

Dr. J.R. Hrynczuk
Gentofte Hospital
Department H
Gentofte (Denmark)

Prof. Dr. H. Hutten
Universität Mainz
Physiologisches Institut
Saarstr. 21
D–6500 Mainz (FRG)

Mr. H. Ijsenbrandt
University of Nijmegen
Department of Physiology
Geert Grooteplein Noord 21
Nijmegen (The Netherlands)

Mr. M. Ikeuchi
Research Institute of Applied Electronics
Hokkaido University
North 12th West 6th
Sapporo (Japan)

Mr. J.R. Ives, Biomed. Ing.
Montreal Neurological Institute
Department of Neurology and Neurosurgery
McGill University
3801 University Ave.
Montreal (Canada)

Mr. R. Jenzer
Jenzer AG Messtechnik
Dorfstr. 41
CH–8954 Geroldswil (Switzerland)

Dr. K. Jeschke
Institut für Leibeserziehung
Innrain 52
A–6020 Innsbruck (Austria)

Mr. J.R. Jonkers
Ziekenhuis St. Annadal
St. Annadal/Maastricht
(The Netherlands)

Mr. H.P. Kaesermann,
Research Associate
Swiss Federal Institute of Technology
Department of Behaviour Sciences
Turnerstr. 1
CH–8006 Zürich (Switzerland)

Dr. P.L. Käfer
Presse- und Informationsdienst
Gloriastr. 35
CH–8006 Zürich (Switzerland)

Dr. H.P. Kägi
Grossmünsterplatz 9
CH–8001 Zürich (Switzerland)

Dr. H. Kaltschmidt
Messerschmidt-Bölkow-Blohm
Unternehmensbereich Apparate und Kybernetik
D–8000 München 80 (FRG)

Dr. D. Kasperkovitz
Philips Research Laboratory
Drof. Holstlaan
Eindhoven (The Netherlands)

Mr. L.T. Kayser, Dipl.-Ing.
Technologieforschung GmbH
Am Bismarckturm 10
D–7000 Stuttgart 1 (FRG)

Mr. J. Kersemakers, El. Eng.
Afd. Cardiologie
Ziekenhuis St. Annadal
St. Annadal/Maastricht
(The Netherlands)

Dr. Ir. H.P. Kimmich
Department of Physiology
University of Nijmegen
Geert Grooteplein Noord 21
Nijmegen (The Netherlands)

Mr. Y.J. Kingma
Erasmus University Rotterdam
Dr. Molewaterplein 50
P.O. Box 1738
Rotterdam (The Netherlands)

Mr. H.J. Klewe, Dipl.-Ing.
Deutsche Forschungs- und Versuchsanstalt für Luft- und Raumfahrt e.V.
Institut für Flugmechanik
D–3300 Braunschweig-Flughafen (FRG)

Dr. J.A.J. Klijn
University of Nijmegen
Department of Neurophysiology
Geert Grooteplein Noord 21
Nijmegen (The Netherlands)

Dr. K. Kocnar
Department of Sports Medicine
University J.E. Purkyne
Pekarska 53
Brno (Czechoslovakia)

Prof. Dr. P. Komi
Kinesiology Laboratory
University of Jyväskylä
SF–40100 Jyväskylä (Finland)

Mr. L.D. Kondratien
Conseil Central des Syndicats soviétique
Moscou (UdSSR)

Dr. H.J. Kongas
University of Oulu
Department of Control and Systems Engineering
Koulukatu 32
SF–90100 Oulu 10 (Finland)

Frau Dr. E. Kovacs
Schweizerische Anstalt für Epileptische
Bleulerstr. 60
CH–8008 Zürich (Switzerland)

Dr. H. Krexa
TU München
Habacherstr. 26
D–8000 München 70 (FRG)

Dr. med. B. Krönig, Ass. Prof.
I. Medizinische Klinik
Langenbeckstr. 1
D–6500 Mainz (FRG)

Prof. G. Küchler
Deutsches Zentralinstitut für Arbeitsmedizin
Nöldnerstr. 40–42
DDR–1134 Berlin-Lichtenberg (GDR)

Mr. H. Kunz, Dipl. El. Ing.
Institute of Applied Physics
Swiss Federal Institute of Technology
CH–8049 Zürich (Switzerland)

Dr. med. Greta Küppers
Hebbelstr. 60
D–5000 Köln 51 (FRG)

Dr. Larcher
ENSEPS
1, rue du Dr Lesavoureux
F–92000 Châtenay-Malabry
(France)

Mr. B. Lavandier
INSERM Unité 37
18, Av. du Doyen Lépine
F–69500 Bron (France)

Mrs. M. Lebert
Sekretariat Turnen und Sport
Plattenstr. 26
CH–8032 Zürich (Switzerland)

Mr. H. Leist
F. Hellige & Co. GmbH
Waldallee 36
D–7800 Freiburg-Lehen (FRG)

Mr. L.E. Lindholm
Department of Applied Electronics
Chalmers University of Technology
Göteborg (Sweden)

Dr. Ch. Lombard, Assistent
Institut für Veterinärphysiologie
Winterthurerstr. 260
CH–8036 Zürich (Switzerland)

Mr. J.A. Maass
Electronic and Bioengineering
Consultant
US Army Materiel Command
Im Mainfeld 23 (15.3)
D–6000 Frankfurt 7 (FRG)

Mr. Magbool
UIT, Union internationale de
Télécommunication
CH–1200 Genève (Switzerland)

Mr. B. Maillet
SFENA
Aérodrome de Villacoublay
BP 59
F–78140 Velizy (France)

Mr. E. Mälkiä, Physiologist
The Rehabilitation Examination
Centre of the Social
Insurance Institute
Peltolantie 3
SF–20720 Turku 72 (Finland)

Mr. M. Marques
Chef Service Biotelemetry
CNRS
Villier-en-Bois
F–79360 Beauvoir (France)

Prof. G. Matsumoto
Research Institute of Applied
Electronics
Hokkaido University
Sapporo (Japan)

Mr. R.V. Mazza
c/o Liebermann
109 W. 28th Street
New York City, NY 10001 (USA)

Dr. J.B. McKinnon, B.Sc., Ph.D.,
M.Inst.P.
Oxford Instruments Co. Ltd.
Osney Mead, Oxford OX2 0DX
(England)

Dr. J.D. Meindl
Stanford University
Stanford, CA 94305 (USA)

Mr. J.G. Meisburger
Ciba-Geigy AG
Geb. K-125/1201
CH–4002 Basel (Switzerland)

Mr. H.D. Melzig, Dipl.-Ing.
Deutsche Forschungs- und
Versuchsanstalt für Luft- und
Raumfahrt e.V.
Institut für Flugmechanik
D–3300 Braunschweig-Flughafen
(FRG)

Dr. V. Mengden
Medizinische Universitätsklinik
D–6500 Mainz (FRG)

Dr. J. Meszaros
Sandoz AG
Gebäude 881/210
CH–4002 Basel (Switzerland)

Dr. K. Meyer-Waarden
Universität
Institut für Biomedizinische
Technik
D–7500 Karlsruhe (FRG)

Mr. R.M. Milton
Chief Research Officer
NEERI, CSIR
P.O. Box 395
ZA–0001 Pretoria (South Africa)

Dr. V. Minarovjech
Clinic of Sport
Medical Faculty Hospital
Mickiewiczova 13
CS–88421 Bratislava
(Czechoslovakia)

Dr. W. Minder
Stadtspital Triemli
Zentrallabor
Birmensdorferstr. 336
CH–8063 Zürich (Switzerland)

Dr. S. Miyashiro
Tokyo Shibaura Electr. Co. Ltd.
Düsseldorf Office
Emmastr. 24–26
D–4000 Düsseldorf (FRG)

Mr. Monbaron
Institut de Physique
Rue A.L. Breguet 1
CH–2000 Neuchâtel (Switzerland)

Dr. K.E. Morander
Selcom AB
Södra Rydsbergsvägen 21
S–44300 Lerum (Sweden)

Mr. J.C. Moreau, Ing.
CNRS-INSERM
5, Square Verrazane
F–77240 Cesson (France)

Mr. B. Müller, Ing. Techn. HTL
Ciba-Geigy AG
K–125/1120
CH–4002 Basel (Switzerland)

Mr. P.A. Neukomm, Dipl. El. Ing.
Swiss Federal Institute of Technology
Laboratory of Biomechanics
Gloriastr. 35
CH–8006 Zürich (Switzerland)

Mr. A. Niewiadomskî, Dipl.-Ing.
Technologische Forschung GmbH
Abteilung Berlin, Institut für Elektronik
D–1000 Berlin 33 (FRG)

Mr. B.M. Nigg
Swiss Federal Institute of Technology
Laboratory of Biomechanics
Plattenstr. 26
CH–8032 Zürich (Switzerland)

Mr. S. Oblin
Université Paris VII
2, Place Jussieu
F–75005 Paris (France)

Mr. K.E.T. Öberg, Res. Eng.
Een Holmgren Ort. AB
Bergsbrunnagatan 1
S–753 23 Uppsala (Sweden)

Mr. W. Oehen, Stud. ETH
Friedheimweg
CH–6353 Weggis (Switzerland)

Mr. D. Olsen
Brain Research Institute
Biology Laboratory UCLA
Los Angeles, CA 90024 (USA)

Mr. J.M. Ory, Ing.
Institut de Génie Biologique et Médical
Parc Robert Bentz
F–54500 Vandoeuvre (France)

Mr. M.A. Ott, Dipl.-Ing.
Institut für Biomedizinische Technik
Keplerstr. 17
D–7000 Stuttgart 1 (FRG)

Dr. J.E. Pearson
Roche Products Ltd.
P.O. Box 8
Welwyn Garden City
Herts. AL7 3AY (England)

Prof. Dr. F. Pellandini
Chaire d'Electronique
Institut de Physique
Rue A.L. Breguet 1
CH–2000 Neuchâtel (Switzerland)

Mr. B. Pfister, Stud. ETH
Kirchfeld 343
CH–6252 Dagmarsellen (Switzerland)

Mr. P. Pinösch
Institut für technologische Physik ETH Zürich
Hönggerberg
Postfach 123
CH–8049 Zürich (Switzerland)

Dr. L. Pircher
FAI, Fliegerärztliches Institut
Bettlistr. 16
CH–8600 Dübendorf (Switzerland)

Mr. J. Place
Mallory Batteries Europe
Kouterveldstr. 15
B–1920 Diegem (Belgium)

Dr. P. Polc
F. Hoffmann-LaRoche & Co. AG
Grenzacherstr. 124
CH–4002 Basel (Switzerland)

Mr. J.J. Potgieter
Chief Research Officer
NEERI, CSIR
P.O. Box 395
ZA–0001 Pretoria (South Africa)

Mr. G. Prins
Nederlands Radar Proefstation
Koningin Astrid Blv. 57
Noordwijk a. Zee (The Netherlands)

Mr. R.D. Rader
Department of Physiology
University of Southern California
815 W. 37th Street
Los Angeles, CA 90007 (USA)

Dr. Dv. Berton A. Rahn
Labor für experimentelle Chirurgie
Schweizerisches Forschungsinstitut
CH–7270 Davos (Switzerland)

Dr. I. Ratov
All Union Scientific Reasearch Institute of Physical Culture of Moscow VNIIFK
ul. Kazakowa 18
Moscow K–64 (USSR)

Dr. J. Rebelle
SFENA
Aérodrome de Villacoublay
BP 59
F–78140 Velizy-Villacoublay (France)

Mr. T. Retterasen, Ing.
Unversity of Oslo
Box 1094
Blindern
Oslo 3 (Norway)

Mr. J. Rintala, Stud. ETH
Riethofstr. 6
CH–8152 Glattbrugg (Switzerland)

Mr. H.F. Rittel
Sportmedizinische Institut TH Aachen
Am Tivoli 11
D–5100 Aachen (FRG)

Mr. H.R. Roman, Med. Electr. Ing.
Orthop. Lab. R.U. Leiden
Clematislaan 69
NL–2407 Oegstgeest (The Netherlands)

Mr. R. Rossi, Project Eng.
OTE Biomedica
Via Caciolle 15
I–50127 Firenze (Italy)

Dr. Roth
Deutsches Zentralinstitut für Arbeitsmedizin
Nöldnerstr. 40–42
DDR–1134 Berlin-Lichtenberg (GDR)

Mr. L.L. Salmon
SA Ets.Salomon et Fils
Chemin de la prairie prol.
F–74003 Annecy (France)

Dr. S. Salmons
University of Birmingham
Department of Anatomy, Medical School
Vincent Drive
Birmingham B15 2TJ (England)

Dr. L. Samek
Benedikt-Kreutz-Rehabilitationszentrum
Südring 15
D–7812 Bad Krozingen (FRG)

Mr. J. Satler
Faculty for Electrical Engineering
Trzaska 25
YU–61000 Ljubljana (Yugoslavia)

Mr. R. Sauter, Stud. ETH
Wuhrstr. 20
c/o S. Enzler
CH–8003 Zürich (Switzerland)

Dr. G. Scarpetta
Cattedra di Elettronica
Instituto Elettronico
Fac. di Ingegneria di Napoli
Via Claudio
I–80125 Napoli (Italy)

Dr. H. Schäfer
Altkönigstr. 17
D–6500 Mainz 32 (FRG)

Mr. H.J. Schmid, Stud. ETH
Baselstr. 35
CH–4132 Muttenz (Switzerland)

Mr. M. Schoenenberger
Schweizerischer Schulrat, Verwaltungskoordination
Revisionsdienst
Pestalozzistr. 24
CH–8028 Zürich (Switzerland)

Dr. W. Schoetensack
Byk Gulden Lomberg Chemische Fabrik
Pharmazeutika
Postfach 500
D–7750 Konstanz (FRG)

Dr. Ch. Scholand
Medizinische Poliklinik der Universität
Östliche Stadtmauerstr. 29
D–8520 Erlangen (FRG)

Dr. B.A. Schölkens
Farbwerke Höchst AG
Abteilung Pharmakologie
D–6230 FEM Hoechst (FRG)

Mr. P. Schweizer, Stud. ETH
Hegaustr. 25
CH–8212 Neuhausen (Switzerland)

Dr. A. Shah
Institute of Applied Physics
Swiss Federal Institute of Technology
CH–8049 Zürich (Switzerland)

Dr. H. Slajmer
Gosposvetskz 27
YU–62000 Maribor (Yugoslavia)

Mr. K.S. Smith, Phys.
Christie Hospital
Withington
Manchester M20 BLR (England)

Mr. W. Steinhagen
Institut für theoretische Elektrotechnik
TH Aachen
D–5100 Aachen (FRG)

Dr. K. Streit
Medical Service of Swiss Railway
Bollwerk 10
CH–3000 Bern (Switzerland)

Mr. S. Sukardi, Stud. ETH
Freudenbergstr. 146
CH–8044 Zürich (Switzerland)

Dr. L.I. Tatarenko
Research Laboratory of Biotelemetry
Picket
Kislovodsk–42 (USSR)

Mr. H. Teulings
Irisstraat 1
St. Oedenrode/Nijnsel
(The Netherlands)

Mr. Ch. Tomei, Phys. Qualifié
CNRS
31, chemin 5 Aiguier
F–13274 Marseille Cédex 2 (France)

Dr. A. Truog
Ciba-Geigy AG
CH–4002 Basel (Switzerland)

Mr. D.J. Twissell
Glaxo Research Ltd.
Greenford Road
Greenford, Middex UB6 0HE (England)

Miss E. Unold
Swiss Federal Institute of Technology
Turnen und Sport/Biomechanik
Plattenstr. 26
CH–8032 Zürich (Switzerland)

Dr. Gerda Verberne
University of Amsterdam
Dier Fysiol. Lab.
Kruislaan 320
Amsterdam (The Netherlands)

Mr. A. Vliegenhart
Electr. Engineering Research
Department
Depex BV
Steenstraat 85
De Bilt (The Netherlands)

Mr. W. Vollenweider, Stud. ETH
Dörfliweg 9
CH–3075 Rüfenacht (Switzerland)

Dr. G.Ch. Voroshkin, Assist. Prof.
Institute of Physical Culture
'G. Dimitrov'
1, Tina Kirkova str.
Sofia–C (Bulgaria)

Dr. J.A. Vos
University of Nijmegen
Department of Physiology
Geert Grooteplein Noord 21
Nijmegen (The Netherlands)

Mr. R.W. Vreeland, Assist.
Dev. Eng.
University of California
Medical Center
Research and Development
Laboratory
San Fransisco, CA 94143 (USA)

Prof. Dr. J. Wartenweiler
Swiss Federal Institute of
Technology
Kurse für Turnen und Sport
Plattenstr. 26
CH–8032 Zürich (Switzerland)

Mr. C. Weller, Biomedical
Engineer
Medical Research Council
Clinical Research Centre
Watford Road
Harrow, Middex HA1 3UJ
(England)

Mr. J. Wespi, Research Associate
Swiss Federal Institute of
Technology
Department of Behaviour Sciences
Turnerstr. 1
CH–8006 Zürich (Switzerland)

Mr. I. White, Pharmacologist
Forschungstechnisches Institut
Wander AG
Monbijoustr. 115
CH–3001 Bern (Switzerland)

Mr. E. Wirz
Ciba-Geigy AG
CH–4002 Basel (Switzerland)

Dr. H. Wolfgang
F. Hoffmann-LaRoche & Co. AG
Grenzacherstr. 124
CH–4002 Basel (Switzerland)

Mr. H.J. Woltring,
Junior Research Fellow
University of Nijmegen
Psych. Laboratory
Erasmuslaan 16
NL–6804 Nijmegen
(The Netherlands)

Mr. C.G. Wouters, El. Eng.
University of Leiden
Suzannalaan 456
NL–2020 Den Haag
(The Netherlands)

Mr. H. Zahnd, Ing. Techn. HTL
Contraves AG
Schaffhauserstr. 580
CH–8052 Zürich (Switzerland)

Dr. H. Zeier
Swiss Federal Institute of
Technology
Department Behaviour Sciences
Turnerstr. 1
CH–8006 Zürich (Switzerland)

Dr. med. R. Zerzawy
Medizinische Poliklinik
Östliche Stadtmauerstr. 29
D–8520 Erlangen (FRG)

Dr. Zimmermann
Deutsche Zentralinstitut für
Arbeitsmedizin
Nöldnerstr. 40–42
DDR–1134 Berlin-Lichtenberg
(GDR)

Author Index

THIS VOLUME MAY CIRCULATE FOR 1 WEEK.
Renewals May Be Made In [illegible] Phone:
x 5300; from [illegible]